SpringerBriefs in Energy

SpringerBriefs in Energy presents concise summaries of cutting-edge research and practical applications in all aspects of Energy. Featuring compact volumes of 50 to 125 pages, the series covers a range of content from professional to academic. Typical topics might include:

- A snapshot of a hot or emerging topic
- A contextual literature review
- A timely report of state-of-the art analytical techniques
- An in-depth case study
- A presentation of core concepts that students must understand in order to make independent contributions.

Briefs allow authors to present their ideas and readers to absorb them with minimal time investment.

Briefs will be published as part of Springer's eBook collection, with millions of users worldwide. In addition, Briefs will be available for individual print and electronic purchase. Briefs are characterized by fast, global electronic dissemination, standard publishing contracts, easy-to-use manuscript preparation and formatting guidelines, and expedited production schedules. We aim for publication 8–12 weeks after acceptance.

Both solicited and unsolicited manuscripts are considered for publication in this series. Briefs can also arise from the scale up of a planned chapter. Instead of simply contributing to an edited volume, the author gets an authored book with the space necessary to provide more data, fundamentals and background on the subject, methodology, future outlook, etc.

SpringerBriefs in Energy contains a distinct subseries focusing on Energy Analysis and edited by Charles Hall, State University of New York. Books for this subseries will emphasize quantitative accounting of energy use and availability, including the potential and limitations of new technologies in terms of energy returned on energy invested.

Thomas L. Brewer

Transforming U.S. Climate Change Policies

2021 and Beyond

Springer

Thomas L. Brewer
School of Business
Georgetown University
Washington, DC, USA

ISSN 2191-5520 ISSN 2191-5539 (electronic)
SpringerBriefs in Energy
ISBN 978-3-030-99715-1 ISBN 978-3-030-99716-8 (eBook)
https://doi.org/10.1007/978-3-030-99716-8

This Springer imprint is published by the registered company Springer Nature Switzerland AG
The registered company address is: Gewerbestrasse 11, 6330 Cham, Switzerland

Preface

This Policy Brief analyzes the period from January 2021 to January 2022, which was an historically significant year for government, business, and the public in the USA. The year marked an extraordinary turning point in U.S. policymaking on climate change issues, when the Biden administration entered office in January 2021. The new administration's rhetoric, executive decisions, and policy proposals differed dramatically from the Trump administration's denial of climate science, support for fossil fuel industries, and disregard of international concerns about the world's changing climate. An understanding of the U.S. policy changes during 2021 inevitably requires an understanding of the domestic political and economic contexts, as well as already-happening changes in the climate, developments in climate science, and evolving technologies for climate change mitigation and adaptation. The Brief thus explains how a pluralistic political system and a diverse economic system shape U.S. national government policies. Governmental institutions share power at the national level, and a multi-level federal system creates overlapping authority among local, state, and national governmental policies. A diverse economic system creates conflicting interests that contribute to regionalized patterns in public opinion and governmental policies. Which policies change—by how much, when, how, and why—depends on votes in legislative bodies, court cases, election outcomes, and industry lobbying activities, as well as policy proposals by the national administration. This Brief focuses on the extent to which the national government's climate policymaking during 2021 featured: an ambitious administration that was determined to advance climate policies along many dimensions; an opposition party that was overwhelmingly against most of the changes; and an institutionalized under-representation in the Senate of support for the administration's agenda. As this Brief was being completed in January 2022, all of the 50 members of the opposition party in the Senate and one of the 50 members of the President's party were preventing passage of core elements of the administration's climate policy initiatives. The impasse was emblematic of the domestic political constraints the administration encountered.

Thomas L. Brewer
Georgetown University
Washington, DC, USA

Contents

1 Climate Change Challenges 1
  1.1   Chapter Overview 1
  1.2   Record Temperatures in the Summer of 2021 1
  1.3   More Extreme Weather Events 2
  1.4   Climate Change in Alaska and the Arctic 4
  1.5   Sources of Climate Change 7
  1.6   Record Carbon Dioxide Concentration Levels 7
  1.7   Implications for Policymaking 8
  Annex 1: Conversions between Metric and U.S. Measurement Units
  [24] 10
  Annex 2: Global Warming Potentials (GWPs) of Gases and Particulate
  Matter 10
  References 12

2 U.S. Policy Agendas 15
  2.1   New Beginnings 15
  2.2   Executive Orders in January 2021 16
  2.3   American Jobs Plan 16
  2.4   Infrastructure Investment and Jobs Act [5] 18
  2.5   Build Back Better Bill [6] 18
  2.6   Executive Order on Federal Government Sustainability [7] 19
  2.7   International Context 19
      2.7.1 The Paris Agreement 19
      2.7.2 Leaders' Summit [9] 20
      2.7.3 G-20 Meeting [10] 20
      2.7.4 COP26 in Glasgow [11–15] 20
      2.7.5 Special Report of the Intergovernmental Panel
           on Climate Change (IPCC) 21
  2.8   Domestic Institutional Constraints 22
  2.9   Public Opinion 23
  2.10 Implications: Presidential Power and Its Limits 25

Annex 1: President Biden's Executive Orders on: (A) Protecting
Public Health and the Environment and Restoring Science to Tackle
the Climate Crisis', (B) 'Tackling the Climate Crisis at Home
and Abroad', and (C) 'Catalyzing Clean Energy Industries and Jobs
Through Federal Sustainability' 26
Annex 2: U.S. Public Opinion Survey Data: Detailed Analyses
of Selected Issues 31
References 33

3 Who Gets What in the Budget 37
   3.1   Overview of the Budget Process 37
   3.2   FY2022 Request to Congress 38
   3.3   Federal Emergency Management Agency (FEMA) [4–6] 41
   3.4   Infrastructure Bill [7] 42
   3.5   Build Back Better Bill [9] 43
   3.6   An Un-representative Senate 44
   3.7   Implications for Climate Change Mitigation and Adaptation 45
   Annex 1: Request for NOAA Appropriations for FY2022 [13] 46
   Annex 2: Build Back Better Budget Changes 48
   Annex 3: Calculations of the 'Social Costs of Carbon' in the Budget
   [14–19] 48
   Annex 4: States' Carbon-Intensity and Representation in the Senate 54
   References 55

4 The Future 57
   4.1   Reframing the Issues 57
   4.2   The U.S. as an International Leader and Laggard 58
   4.3   Pricing Carbon 58
   4.4   Institutional Constraints 59
   4.5   The Generation Gap in Public Opinion 59
   References 61

List of Figures

Fig. 1.1   Black carbon deposits on Greenland's Glaciers [13]   5
Fig. 1.2   Extent of Arctic Sea Ice during November–March [15]   6
Fig. 1.3   Shares and Trends in U.S. Economic Sectors' emissions (does
           not include BC particulate emissions) [21] (million metric
           tonnes of CO_2e)   9
Fig. 1.4   Annual and seasonal concentration levels of carbon dioxide
           (1958–2021) [23]   10
Fig. 3.1   Map of areas of the U.S that are most exposed to flooding [8]   44

List of Tables

Table 1.1   Types of emissions and their sources in economic
sectors [20] 8

Table 1.2   Metric conversion factors 11

Table 1.3   Global Warming Potentials (GWPs) for selected greenhouse
gasses and particulate matter [20, 25] 11

Table 2.1   Opinions in the U.S. about climate change
before and after the summer of 2021 [21] 23

Table 2.2   U.S. Demographic group differences in public perceptions
of climate change [22] 24

Table 2.3   Public perceptions of major threats: international
comparisons of the US and 13 other countries [23] 25

Table 2.4   Comparison of results using 'Climate Change' or 'Global
Warming' in survey questions [24] 31

Table 2.5   Basic beliefs: comparisons among democrats, independents
and republications [25] 32

Table 2.6   Core policy preferences: comparisons among democrats,
independents and republicans [26] 33

Table 3.1   The administration's FY2022 request for 'Tackling
the Climate Crisis' [1] 39

Table 3.2   Energy department FY2022 budget request [2] 40

Table 3.3   Amounts for Selected Climate-Related Items Programs
in 2021 Infrastructure Legislation [7] 43

Table 3.4   Per capita carbon dioxide emissions of ten highest-ranking
states [11] 45

Table 3.5   Proposed climate change appropriations in the build back
better bill [9] 49

Table 3.6   Variations in estimated economic values imputed to a tonne
of carbon dioxide [15] 54

Table 3.7 Carbon intensity of U.S. states (CO_2e emissions per million
          USD million GDP) [20] 55
Table 4.1 Partisan and age differences in climate change attitudes -
          reducing effects of climate change should be a top priority [9] ... 60

List of Boxes

Box 1.1   U.S. Summer Temperatures (2021) [2] 2

Box 1.2   Droughts, Fires and Floods [3] 2

Box 1.3   Conditions and Trends in the Arctic [8–11] 4

Box 2.1   Highlights of the American Jobs Plan [4] 17

Box 3.1   Highlights of the Energy Department's Justification of Its
FY2022 Request [2] 38

Chapter 1
Climate Change Challenges

1.1 Chapter Overview

This introductory chapter presents a series of short reminders of how the trends and patterns of recent years have led to widespread recognition that climate change has reached crisis levels and that more serious mitigation and adaptation measures are urgently needed. There is an emphasis on temperatures and emissions in 2020 and 2021 within long-term trends. The analysis puts the issues for the U.S. in an international comparative context as well as a domestic U.S. context. The Arctic is highlighted because it has a special place as a region that has been warming at four times the global average rate and because the U.S. has significant interests and responsibilities there, including membership in the Arctic Council.

The chapter is also an introduction for readers who are not already conversant with some of the key concepts and data of climate science. There are technical annexes that provide more detailed information about selected topics: conversion ratios for metric and U.S. measurement units; and global warming potentials (GWPs) of greenhouse gases and particulate matter.

1.2 Record Temperatures in the Summer of 2021

The summer of 2021 tied with 1936 as the warmest U.S. summer on record [1]. This and other highlights of the summer heat are in Box 1.1.

© The Author(s), under exclusive license to Springer Nature Switzerland AG 2022   1
T. L. Brewer, *Transforming U.S. Climate Change Policies*,
SpringerBriefs in Energy, https://doi.org/10.1007/978-3-030-99716-8_1

Box 1.1: U.S. Summer Temperatures (2021) [2]

During meteorological summer (June–August), the average temperature for the Lower 48 states was 74.0 °F, 2.6 °F above average, nominally eclipsing the extreme heat of the Dust Bowl in 1936 by about 0.01°F. A record 18.4% of the contiguous U.S. experienced record-warm temperatures for this season. For August, the contiguous U.S. average temperature was also 74.0 °F, 1.9 °F above the twentieth century average and ranked as the 14th-warmest August on record. For the year to date, the contiguous U.S. temperature was 55.6, 1.8 °F above the twentieth century average, ranking 13th warmest in the January-August record.

...

Summer temperatures were above average to record warmest from the West Coast to the Great Lakes and into the Northeast as well as across portions of the Mid-Atlantic and Gulf Coast. California, Nevada, Utah, Oregon and Idaho each reported their warmest June–August on record.

1.3   More Extreme Weather Events

Box 1.2: Droughts, Fires and Floods [3]

According to the August 31 [2021] U.S. Drought Monitor report, approximately 46.6% of the contiguous U.S. was in drought, slightly more than the coverage at the beginning of August. Drought conditions expanded or intensified across the Northern Tier, the Pacific Northwest and portions of California. Drought coverage and/or intensity lessened across parts of the Four Corners region, the Midwest, Hawaii and Puerto Rico and was eliminated in Alaska.

...

Wildfires continued to spread across the western U.S. during August as the Dixie Fire in north-central California became the second-largest fire in the state's history. The Caldor Fire also in California grew rapidly during August, threatening South Lake Tahoe communities. Air quality remained a concern across the U.S. as ash and fine particulates from the many wildfires obscured the skies.

...

The *summer precipitation* total across the Lower 48 states was 9.48 in., equal to 1.16 in. above average, ranking eighth wettest in the historical record. The

August precipitation total for the contiguous U.S. was 3.09 in., 0.47 in. above average, ranking 14th wettest in the 127-year period of record. The year-to-date precipitation total across the contiguous U.S. was 21.19 in., or 0.48 in. above the long-term average, ranking in the middle third of the January-August record.

...

Devastating flash flooding and fatalities resulted from multiple events during August including Tropical Storm Fred in western North Carolina, convective flooding from a complex of storms across middle Tennessee, Hurricane Ida across Louisiana and portions of the Northeast in early September and from Tropical Storm Henri, also across parts of the Northeast. With 35 fatalities accounted for during August, it was the deadliest month for flooding across the U.S. since Hurricane Harvey in 2017.

Extreme weather events continued after the end of the summer. An unprecedented drought in Colorado late in the year—when and where there would normally be snow that time of the year—fed wild fires that destroyed hundreds of homes near Denver [4].

Also, in December, the most extensive and destructive tornadoes occurred in recorded U.S. history, when there was a winter-time, record-warm series of days [5–7]. More than 30 tornadoes created a path of death and destruction over 200 miles long in five states. More than 70 people were killed, and preliminary estimates of the economic damages were on the order of hundreds of billions of dollars.

Because the accumulated evidence from climate science research has not been as extensive for tornadoes as for other kinds of extreme weather events, there were not yet widely accepted answers among climate scientists in 2021 to key questions about causal connections between climate change and tornadoes. In particular, it was not yet clear whether the *frequency* or *intensity* of tornadoes is affected by climate change.

However, the occurrence of the 2021 December tornadoes during a period of *extraordinarily warm winter weather* is consistent with the physics of tornadoes; they occur when warm air passes over major storm fronts—*typically in the spring tornado season*. Thus, the demonstrably higher average winter temperatures associated with global warming can logically be expected to lead to more winter tornadoes. Whether such a pattern will be supported by climate scientists' empirical observations remains to be determined; whether there will be a change in *yearly totals* also remains to be determined.

Other important questions about trends and patterns in U.S. tornadoes concerns shifts in their *location*. These questions, though, were not only about whether climate change was contributing to the observed shift eastward in the relative numbers of U.S. tornadoes from the southwestern and western states of Texas, Oklahoma,

Kansas and Nebraska. There has been a shift to southern and central states such as Arkansas, Illinois, Kentucky, Missouri, and Tennessee, where there were deaths from the December 2021 tornadoes. The questions were also about demographic differences among the affected states—especially how much greater the damages could be in the more densely populated states in the latter group. Of course, this is explicitly a socio-economic issue, not a climate science issue.

There was one more high temperature record during the winter of 2021: on the last day of the year, there were temperatures in Alaska that were higher than the temperatures in southern California on the same day.

1.4   Climate Change in Alaska and the Arctic

The Arctic is sometimes symbolically referred to as the climate change equivalent of the 'canary in the mine'—an analogy with the use of the deaths of caged canaries in coal mines to indicate dangerous levels of carbon monoxide for the human workers in them. Because of the presence of Alaska in the Arctic region and thus U.S. membership on the Arctic Council with seven other Arctic countries, the Arctic is a region of special interest for this Brief. As Box 1.3 documents, recent developments in the Arctic region are having significant impacts in the other 49 U.S. states and the rest of the world.

Box 1.3: Conditions and Trends in the Arctic [8–11]

Global warming is real, rapid, and relentless. According to NASA [the US National Aeronautics and Space Administration], Arctic Ocean sea ice extent in September, the time of year when it reaches its minimum, has declined more than 13% per decade since 1979. Moreover, a series of reports and findings in the last quarter of 2020 indicate a landscape undergoing dramatic change. The National Snow and Ice Data Center (NSIDC) announced that 2020's September Arctic sea ice minimum was the second lowest recorded when compared to the lowest extent observed in September 2012. More disturbingly NSIDC also reports October sea ice satellite data indicated the lowest recorded sea ice coverage for the month, and significant portions of the Arctic Ocean remained ice free in the month of November. And as the year [2020] drew to a close, the National Oceanic and Atmospheric Administration (NOAA) released in December its 15th annual Arctic Report Card. NOAA's report paints a stark picture: The Arctic continues to warm, melt, thaw, green, erode, and dry at a pace far quicker than previously forecast.

In the Arctic, black carbon and methane, as well as carbon dioxide, are all major climate change agents. Climate scientists have long known that black carbon is an important climate change forcing agent; the first Assessment Review in 1990 by the International Panel on Climate Change (IPCC) noted its existence as a climate change forcing agent [8]. However, because black carbon is not a gas, it has often been left out of policymaking discourse about the sources of global warming [9, 10]. One of the frontiers of climate change analysis for many years has been the integration of black carbon data and related equations in mathematical models that describe patterns and trends in climate change—and project them into the future [11].

The photograph in Fig. 1.1 reveals the extent of black carbon (BC) deposits on the snow and ice surfaces in the Arctic—particularly in Greenland. The BC on Greenland is a major cause of melting glaciers, which discharge enormous volumes of fresh water into the nearby Arctic Sea—and that, in turn, changes Atlantic ocean salinity, temperature and currents; Greenland's glacial melt also contributes significantly to global sea-level rise [12]. All of these effects are felt in the U.S., especially along the Atlantic coast from Maine to Florida.

Black carbon deposits on Arctic Sea ice have different consequences from those on Greenland's glaciers because ice floating in the sea displaces its water-equivalent volume. Nevertheless, the extent of sea ice has decreased and the possibilities for maritime shipping traffic to traverse the Arctic have increased.

Figure 1.2 shows the substantial decrease of the extent of Arctic Sea ice in recent years during the period November to March, with its peak in the spring and its annual low point in the fall. The unusually low extent in 2020–2021 is evident in the blue

Fig. 1.1 Black carbon deposits on Greenland's Glaciers [13]

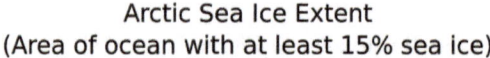

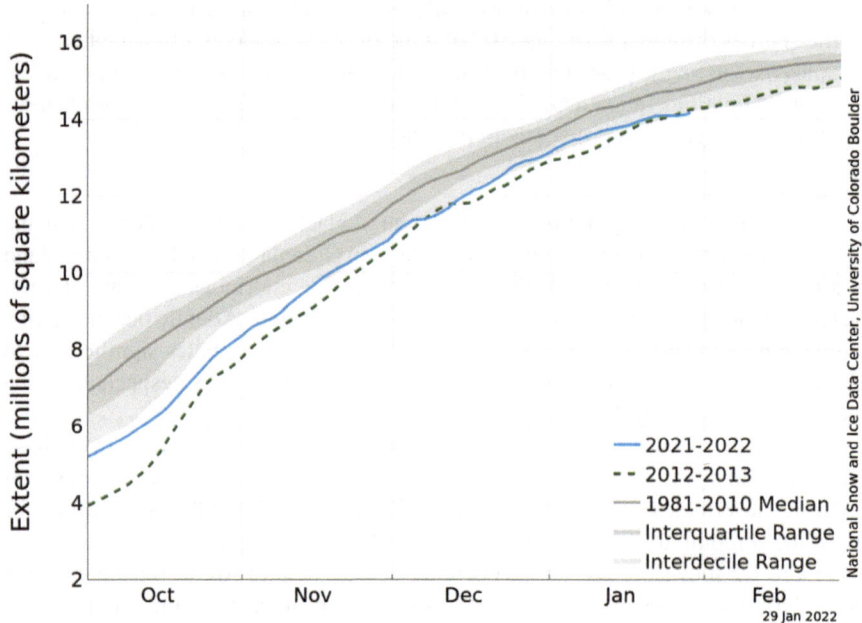

Fig. 1.2 Extent of Arctic Sea Ice during November–March [15]

line at the bottom. Concomitant increases in the volumes of cruise ship and cargo ship traffic are already evident. Increased Arctic ship traffic adds to the black carbon burden in the region—which is the result of a combination of local sources such as wood stoves and quite distant sources such as coal-fired electric power plants in China [14].

Black carbon is not the only climate forcing agent in the Arctic region. Carbon dioxide and methane are also present. As in other regions of the world, carbon dioxide is present as a globally mixed, long-lived greenhouse gas, and it thus has continuing warming effects year after year. Its current effects on Arctic ice melt, however, may be less than those of black carbon.

Methane emissions and leaks in the Arctic region pose yet other issues. Like black carbon, methane is highly potent relative to carbon dioxide—as much as 84 times more potent per tonne at 20 years. Methane leaks from oil wells are an important contributor to Arctic warming [16]. Another source of Arctic methane is the leaks that are already occurring from beneath the melting Arctic tundra, including in human-occupied areas of Alaska in the US—and similarly in northern Canada and Russia [17]. Yet another source of methane leaks is the potential of the enormous reserves trapped under the Arctic Sea, which has a potential total global warming impact that would be greater than all of the greenhouse gases released since the industrial revolution in the eighteenth and nineteenth centuries [18]. The methane reserves under the tundra and the sea in the Arctic region could cause 'tipping points'—beyond which global climate change would be irreversible [19].

Carbon dioxide, methane and black carbon are not only significant climate change agents in the Arctic; they are present around the world, in many economic sectors and in urban and rural areas as well, including the US. Indeed, all global climate-forcing agents contributed to the extraordinary summer of 2021 in the US.

1.5 Sources of Climate Change

There are two linked levels of the 'sources' of climate change: (a) the kinds of chemicals that are emitted and (b) the sources of the emissions. For instance, as just noted above for the Arctic, there are carbon dioxide, black carbon and methane emissions, in terms of chemistry; and there are also oil production facilities and electricity production facilities as well as local and international shipping, in terms of economic sectors and their technologies, that contribute to global warming. More generally, it is well known that the energy sector and the transportation sector are major sources of greenhouse gases and particulate matter.

As Table 1.1 indicates, there are many types of chemicals in global warming emissions and in economic sectors that produce the emissions. Among the many kinds of emissions in the table, the four that are the biggest contributors are carbon dioxide and nitrogen dioxide, as well as black carbon and methane. The table presents brief summaries of the emissions and the industries that emit them—which are considered in greater detail in chapter 5 on industry-specific issues, including their related technological issues and government policy issues.

Figure 1.3 indicates the overall trends in the emissions of the principal economic sectors in the U.S. It is apparent there that the electric power sector is the only one that experienced a notable decline during the period 2005–2019, when natural gas increasingly became a cheaper alternative to coal. Because of the covid pandemic, 2020 and 2021 were of course aberrant years in the long-term trend in emissions.

During the covid pandemic years 2020–2021, there was a dip in emissions during 2020 and then a resumption of upward trends in 2021 [22].

1.6 Record Carbon Dioxide Concentration Levels

Cumulative annual greenhouse gas emissions are reflected in the atmospheric concentration levels of carbon dioxide of course. The long-term trend and the annual seasonal variations in the carbon dioxide concentration levels and Keeling curve are depicted in Fig. 1.2. The curve is notable for its long-term continuously increasing carbon dioxide concentration level—at an increasing rate over time—and its seasonal fluctuations around the annual means. The seasonal pattern is the result of the relative

Table 1.1 Types of emissions and their sources in economic sectors [20]

Emissions	Key Features of the Emissions and Their Sectoral Sources
Short-lived	
Black carbon (BC)	Particulate matter ($PM_{1.0}$), not a gas; one of the three leading climate change forcing agents; also causes respiratory, cardiovascular and other diseases; damage to plants reduces food production; diesel engines are a major source
Methane (CH_4)	Highly potent GHG with 20-year global warming potential (GWP) 84 times carbon dioxide; leaks from production, distribution and usage; atmospheric life of a decade; precursor to ozone
Ozone (O_3)	Secondary pollutant resulting from NO_x and VOCs, with health effects, as well as global warming effects
Aerosol clouds	Airplane contrails are a common form
Organic carbon (OC)	Climate change coolant; co-pollutant with BC, but in smaller volumes than BC as an emission from diesel engines
Long-lived	
Carbon dioxide (CO_2)	Leading climate change forcing agent, emitted by all transportation modes' internal combustion engines and fossil fuel electricity generating facilities
Nitrous oxides (NO_x)	Long-lived with a GWP that is hundreds of times greater than CO_2; also a precursor to ozone
Sulphur oxides (SO_x)	Sulfur dioxide (SO_2) affects climate change as a *coolant*; damages human health and also indirectly contributes to the formation of acid rain
Sulfur hexafluoride (SF_6)	Long-lived (more than 3000 years) and potent (GWP more than 17,000 at 20 years and more than 23,000 at 100 years)
Some short-lived, some long-lived	
Fluorocarbons Hydroflourocarbons Perflourocarbons (CFCs, HFCs, HCFCs, PFCs)	Extremely potent climate change agents, some with GWPs on the order of thousands to tens of thousands; highly variable atmospheric lifetimes from a few days to thousands of years

high rate of carbon dioxide absorption by trees in the summer and low rate in the winter (Fig. 1.4).

1.7 Implications for Policymaking

In sum, indications of record-breaking climate conditions and their impacts have reinforced the widespread recognition among climate scientists that there is now a climate crisis and that much more serious policy measures are needed to address it. The relevant policies include both mitigation polices to reduce emissions and adaptation policies to reduce the damages.

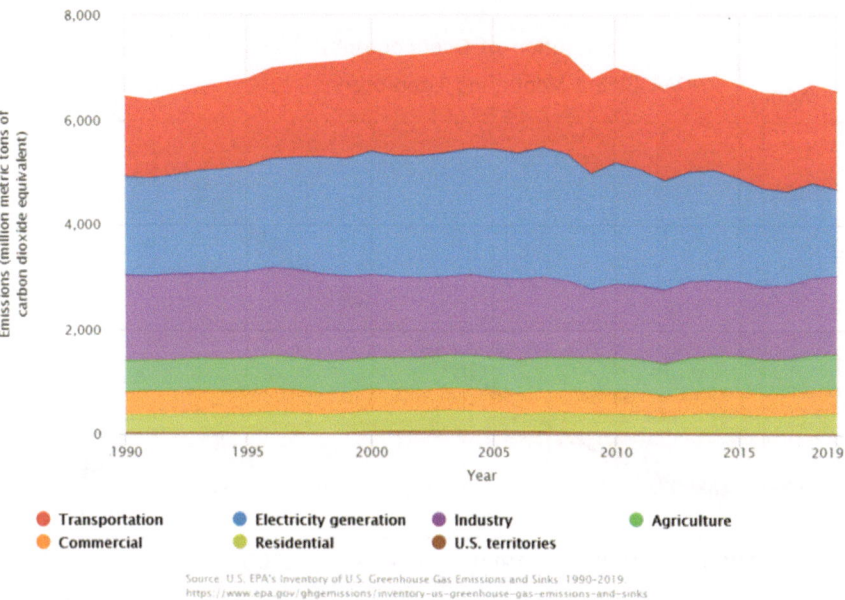

U.S. Greenhouse Gas Emissions by Economic Sector, 1990–2019

Fig. 1.3 Shares and Trends in U.S. Economic Sectors' emissions (does not include BC particulate emissions) [21] (million metric tonnes of CO_2e)

The mitigation and adaptation policy issues and options are analyzed in the next chapter in the context of the Biden administration's intention to transform both mitigation and adaptation policies. A key feature of the economics and politics of the issues is that they vary across regions, with some regions being especially concerned about mitigation policies and some regions being especially concerned about adaptation policies. As for the former, some economies are relatively carbon-intensive in their emissions; fossil fuel producing regions are obviously especially significant in this respect. At the same time, some regions are more vulnerable to sea level rise –obviously, the Atlantic, Gulf of Mexico and Pacific coasts. These regionalized differences in the impacts of the changing climate and the efforts to change climate policies contribute directly to the highly conflictual policymaking processes.

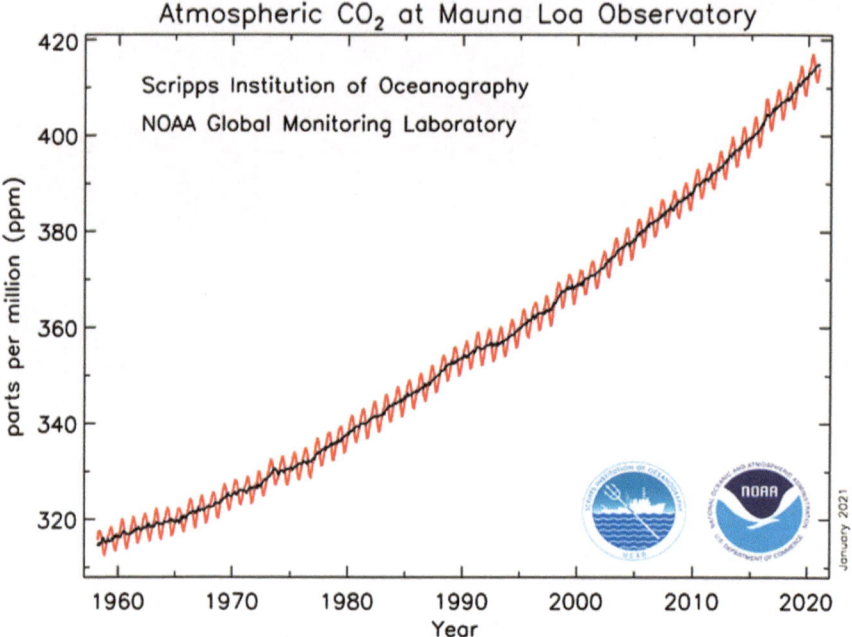

Fig. 1.4 Annual and seasonal concentration levels of carbon dioxide (1958–2021) [23]

Annex 1: Conversions between Metric and U.S. Measurement Units [24]

Most data in the Brief are presented in metric units because they are the usual standard among climate scientists, especially in documents for international audiences. The conversion table (Table 1.2) is presented for the convenience of US readers.

Annex 2: Global Warming Potentials (GWPs) of Gases and Particulate Matter

Global Warming Potential (GWP) is the most widely used metric to indicate the relative potency per tonne of individual types of emissions, compared with carbon dioxide. There are two common shortcomings concerning black carbon (BC) particulate matter. One is that BC is typically excluded from lists of climate change pollutants because it is not a gas, and there is a strong tradition of listing the GWPs of gasses only.

Another issue arises because BC—like methane—is short-lived compared with carbon dioxide. Again, there is a strong tradition of using a 100-year time frame since

Table 1.2 Metric conversion factors

Weights
1 pound = 0.454 kg = 16 oz
1 kg = 2.205 pounds = 35.27 oz
1 short ton = 0.9072 metric tons = 2000 pounds
Volumes
1 ft^3 = 0.02832 cubic meters = 28.3168 L
1 m^3 = 35.315 cubic feet = 1000 L
1 U.S. gallon = 3.78541 L = 0.03175 barrels = 0.02381 barrels petroleum
1 L = 0.2642 U.S. gallons = 0.0084 barrels = 0.0063 barrels petroleum
1 barrel = 31.5 U.S. gallons = 119 L = 0.75 barrels petroleum
1 barrel petroleum = 42 U.S. gallons = 159 L
Distances and areas
1 mile = 1.609 km = 5280 ft
1 km = 0.6214 miles = 3280.84 ft
1 square mile = 2.590 km^2 = 640 acres
1 km^2 = 0.386 square miles = 100 ha
1 acre = 43,560 ft^2 = 0.4047 ha = 4047 m^2

the average atmospheric lifetime of carbon dioxide is approximately 100 years. A 20-year lifetime, which is sometimes included in GWP tables, is particularly relevant in the current era, when there is a consensus among climate scientists that there must be significant reductions in global emissions within about a decade in order to avoid much greater catastrophic consequences.

Some models of future emissions paths that are needed in order to avoid exceeding the Paris Agreement target of limiting global temperatures to less than 1.5 °C or 2.0 °C by 2050 indicate that there must be significant cuts in both BC and methane emissions as well as carbon dioxide emissions.

Table 1.3 thus includes black carbon along with carbon dioxide, methane and nitrous oxide—and it presents both 20-year and 100-year comparisons.

Table 1.3 Global Warming Potentials (GWPs) for selected greenhouse gasses and particulate matter [20, 25]

Chemical name	GWP 20 years	GWP 100 years
Carbon dioxide	1	1
Methane	84	28
Nitrous oxide	264	265
Black carbon	3200	900

References

1. U.S. National Oceanic and Atmospheric Administration. U.S. (2021). *Climate Summary for August and summer 2021*. September 27, 2021. https://www.climate.gov/news-features/unders tanding-climate/us-climate-summary-august-and-summer-2021. Accessed 28 January 2022.
2. U.S. National Oceanic and Atmospheric Administration. (2021). *National Centers for Environmental Information*. National Climate Report—August 2021. Summer Highlights. https://www.ncdc.noaa.gov/sotc/national/202108. Accessed January 2022.
3. U.S. National Oceanic and Atmospheric Administration. (2021). *U.S. Drought Monitor Report Update for August 31, 2021*. https://www.ncei.noaa.gov/news/us-drought-monitor-update-aug ust-31-2021. Accessed January 28, 2022
4. U.S. National Oceanic and Atmospheric Administration. (2022). *Billion-Dollar Weather and Climate Disasters*. https://www.ncdc.noaa.gov/billions/events/US/202. Accessed January 28, 2022
5. U.S. National Oceanic and Atmospheric Administration. (2021*). The December 2021 tornado outbreak explained*. December 20, 2021. https://www.noaa.gov/news/december-2021-tornado-outbreak-explained. Accessed January 28, 2022
6. U.S. National Weather Service. (2021). *NWS Storm Damage Summaries*. December 10–11, 2021. https://www.weather.gov/crh/dec112021. Accessed 28 January 2022.
7. U.S., NOAA National Severe Storms Lab. (2022). *Tornado Basics*. https://www.nssl.noaa.gov/ education/svrwx101/tornadoes/. Accessed January 28, 2022.
8. Sfraga, M. (2021). Navigating the Arctic's 7Cs, US Coast Guard. *Proceedings of the Marine Safety & Security Council, 78*(1), 6.
9. Brewer, T. (2019). The Arctic Ocean's melting ice: Institutions and policies to manage black carbon. In P. G. Harris (Ed.), *Climate change and ocean governance*. Cambridge University Press.
10. Brewer, T. (2016). *Proposal for an Arctic Black Carbon (ABC) Agreement*. Presentation at workshop on Marine Black Carbon Emissions: BC Control Strategies, sponsored by the International Council on Clean Transport (ICCT) in collaboration with Environment and Climate Change Canada, Vancouver, September 2016.
11. University Corporation for Atmospheric Research (UCAR). (2022). *Greenland's Ice is Melting.*https://scied.ucar.edu/learning-zone/climate-change-impacts/greenlands-ice-melting. Accessed January 29, 2022.
12. Voosen, P. (2020). Seas are rising faster than eve. *Science*, November 18, 2020. https://www.sciencemag.org/news/2020/11/seas-are-rising-faster-ever. Accessed on January 26, 2021
13. Cho, R. (2016). The damaging effects of black carbon, state of the planet. *Columbia University, Earth Institute*, March 22, 2016. Photo by Wing-Chi Poon. https://blogs.ei.columbia.edu/2016/ 03/22/the-damaging-effects-of-black-carbon/. Accessed March 25, 2021.
14. Yamineva, Y. (2020). Reducing China's Black Carbon Emissions: An Arctic Dimension. *The Arctic Institute*. April 14, 2020. https://www.thearcticinstitute.org/reducing-china-black-car bon-emissions-arctic-dimension/. Accessed January 29, 2022.
15. U.S. National Snow and Ice Data Center. (2022). *Arctic Sea Ice Extent*. January 29, 2022. http:// nsidc.org/arcticseaicenews/. Accessed January 29, 2022.
16. Arctic Council. (2022). *Significant Economic and Environmental Gains Can Be Achieved by Applying Best Available Technology In The Oil Sector in the Arctic*. https://arctic-council.org/ news/best-available-technology-in-the-oil-sector-in-the-arctic/. Accessed January 29, 2022.
17. Foitzheim, N., Majka, J., & Zastrozhnov, D. (2021). Methane release from carbonate rock formations in the Siberian permafrost area during and after the 2020 heat wave. *Proceedings of the National Academy of Sciences (PNAS), 118*(32), e2107632118. https://doi.org/10.1073/ pnas.2107632118. Accessed January 29, 2022.
18. Science Daily. (2021). *Arctic methane release due to melting ice is likely to happen again*. March 22, 2021. https://www.sciencedaily.com/releases/2021/03/210322135221.htm. Accessed January 29, 2022.

19. More than methane reserves under the tundra and the sea in the Arctic region could cause 'tipping points' beyond which global climate change would be irreversible.
20. Intergovernmental Panel on Climate Change (IPCC). (2021). *Climate Change 2021: The Physical Science Basis.* Contribution of Working Group I to the Sixth Assessment Report of the Intergovernmental Panel on Climate Change; chapters 2, 3, 5, 6, 7, 12. Cambridge, UK: Cambridge University Press.
21. U.S. Environmental Protection Agency (EPA). (2021). Inventory of U.S. Greenhouse Gas Emissions and Sinks, 1990–1999. Accessed December 15, 2021.
22. Rhodium Group, Preliminary US Greenhouse Gas Emissions Estimates for 2020. https://rhg.com/research/preliminary-us-emissions-2020/. Accessed on January 26, 2021.
23. U.S. National Oceanic and Atmospheric Administration (NOAA). (2021). *Assessing the U.S. Climate in August 2021.* www.ncei.noaa.gov/news/national-climate-202108. Accessed September 10, 2021.
24. U.S. Environmental Protection Agency (EPA). (2022). *Greenhouse Gas Equivalencies Calculator.* https://www.epa.gov/energy/greenhouse-gas-equivalencies-calculator. Accessed January 27, 2022.
25. Bond, T., et al. (2013). Bounding the role of black carbon in the climate system: A scientific assessment. *Journal of Geophysical Research: Atmospheres,118,* 5380–5552.

Chapter 2
U.S. Policy Agendas

2.1 New Beginnings

Climate related events, science and policymaking during 2021 transformed the U.S. climate change mitigation and adaptation agendas:

- A series of extreme weather events with unexpectedly severe impacts were substantially attributed to global warming, and they thus provided evidence that the effects of global warming were already happening.
- A report by the UN Intergovernmental Panel on Climate Change (IPCC) contained alarming evidence and projections that underscored the urgent need to take action in order to avoid the most catastrophic impacts of global warming.
- A change in the US national government from the previous administration—which was dominated by deniers of the science and supporters of carbon-intensive industries—to an administration that was calling for a wide range of policy initiatives to strengthen climate change mitigation and adaptation measures.

The Biden administration developed an ambitious, wide-ranging agenda of policy changes they planned for the first year. The changes included: reducing subsidies to oil and gas companies while increasing subsidies for sustainable alternatives to fossil fuels; changing the methods of computing the 'social cost of carbon' and thus the economic cost–benefit assessments of policies; insuring that national government infrastructure projects reduce climate change emissions; preparing a National Intelligence Estimate on the effects of climate change on national security and economic security; advancing a climate action agenda in international venues, such as the Conference of the Parties (COP) of the UN Framework Convention on Climate Change (FCCC), the Group of Seven (G7), the Group of Twenty (G20), and international programs concerning energy, aviation, shipping, the Arctic, the oceans, sustainable development and migration.

© The Author(s), under exclusive license to Springer Nature Switzerland AG 2022 15
T. L. Brewer, *Transforming U.S. Climate Change Policies*,
SpringerBriefs in Energy, https://doi.org/10.1007/978-3-030-99716-8_2

2.2   Executive Orders in January 2021

On his first day in office, 20 January 2021, President Biden initiated a formal legal process to re-enter the Paris Agreement [1]. That action reversed the decision of the previous Trump administration four years earlier to withdraw from the Agreement. On the same day, President Biden also issued an 'Executive Order on Protecting Public Health and the Environment and Restoring Science to Tackle the Climate Crisis'[2]. In it, he stated that:

> [T]he Federal Government must be guided by the best science and be protected by processes that ensure the integrity of Federal decision-making. It is therefore, the policy of my Administration to listen to the science to improve public health and protect our environment.

A week later the President issued a lengthy and detailed 'Executive Order on Tackling the Climate Crisis at Home and Abroad' [3]. That document began with the international context of U.S. climate policymaking and a reiteration of a commitment to use science in addressing climate change issues:

> The United States and the world face a profound climate crisis. We have a narrow moment to pursue action at home and abroad in order to avoid the most catastrophic impacts of that crisis and to seize the opportunity that tackling climate change presents. Domestic action must go hand in hand with United States international leadership, aimed at significantly enhancing global action. Together, we must listen to science and meet the moment.

Excerpts from the Executive Orders are in this chapter's Annex 1. These early high-profile actions marked the beginning of numerous initiatives to change U.S. climate policy over the following months and years. Some had been successful, others not, and yet others pending, as of early 2022. In any case, it was an unprecedented year for U.S. policymaking, as the following specifics make clear.

2.3   American Jobs Plan

On 31 March, 2021, a policy initiative marked major changes in the issue contexts in which climate policies were to be considered. In particular, the president announced The American Jobs Plan [4]. The plan set forth a wide range of legislative priorities, with total estimated expenditures of approximately $2 trillion over eight years (2022–2029), including approximately $700 billion or about one-third of the total for explicit climate change, energy efficiency and/or other climate-friendly energy programs. The costs for the entire package of programs were to be offset by increases in corporate taxes and reductions in government subsidies for fossil fuel production. Since many of the proposals required legislation to establish new programs, they needed congressional approval; in any case, the expenditure proposals needed congressional approval through its budgeting processes. The plan's highlights in Box 2.1 are suggestive of the diversity and context of the proposals.

Box 2.1 Highlights of the American Jobs Plan [4]

The plan is responsive to the need to 'improve infrastructure resilience'. It notes that in the US during 2020 there were '22 separate billion-dollar weather and climate disasters, costing $95 billion in damages to homes, businesses, and public infrastructure. In Louisiana, Hurricane Laura caused $19 billion of damage in broken water systems and a severely damaged electrical grid'.

A trans-sectoral R&D program includes:

[Establish an] ARPA-C to develop new methods for reducing emissions and building climate resilience, as well as expanding across-the-board funding for climate research. In addition to a $5 billion increase in funding for other climate-focused research, [the] plan will invest $15 billion in demonstration projects for climate R&D priorities, including utility-scale energy storage, carbon capture and storage, hydrogen, advanced nuclear, rare earth element separations, floating offshore wind, biofuel/bioproducts, quantum computing, and electric vehicles ....

Transportation programs include:

[M]odernize the bridges, highways, roads, and main streets that are in most critical need of repair. This includes funding to improve air quality, limit greenhouse gas emissions....

Modernize public transit. ... Double federal funding for public transit ....

[B]uild a national network of 500,000 [electric vehicle] chargers by 2030 .... [R]eplace 50,000 diesel transit vehicles and electrify at least 20% of our yellow school bus fleet....

Energy:

[In order to be on a] path to achieving 100% carbon-free electricity by 2035 [there are proposals for] ... a ten-year extension and phase down of an expanded direct-pay investment tax credit and production tax credit for clean energy generation and storage ... purchasing 24/7 clean power for federal buildings ... and continuing to leverage the carbon pollution-free energy provided by existing sources like nuclear and hydropower....

Buildings:

[Invest] $213 billion to produce, preserve, and retrofit more than two million affordable and sustainable places to live.

[U]pgrade homes through block grant programs, the Weatherization Assistance Program, and by extending and expanding home and commercial efficiency tax credits.

Industry:

[The federal government's] purchasing power can be used to drive innovation and clean energy production, as well as to support high quality jobs. To meet the ... goals of achieving net-zero emissions by 2050, the United States will need more electric vehicles, charging ports, and electric heat pumps for residential heating and commercial buildings ... as well as critical technologies like advanced nuclear reactors and fuel, here at home through a $46 billion investment in federal [government] buying power ....

There was much public support in general, as well as support among many segments of U.S. business, for more U.S. government action to mitigate climate change; however, there was also still much opposition from some sectors of the economy, some regions of the country and some ideologically opposed groups, as well as a substantial majority of the members of the opposition party in Congress. There were therefore many months and years of conflict ahead as the administration pushed forward with its policy change proposals with a mix of support and opposition in congress, the courts, state and local governments, business interest groups and the public.

When the administration's proposals faced opposition in Congress, the President divided the American Jobs Plan into two separate proposals—one, the 'infrastructure' bill that included climate-related programs, and the other, the 'Build Back Better' bill that was a mixture of programs concerning health, education and other 'social' issues, as well as climate change.

2.4   Infrastructure Investment and Jobs Act [5]

A frequently stated rationale for many of the proposals in the infrastructure plan is the achievement of a 'carbon free economy by 2050.' The plan also mentions de-carbonizing industry and carbon-free electricity as goals. The specifics of the climate-related proposals can be grouped in economic sectors for identifying the sources of emissions, such as transportation, energy, buildings and industry. The budget details of the act are analyzed in Chap. 3, which is focused on the fates of several Biden climate initiatives in the Congressional budgeting process.

2.5   Build Back Better Bill [6]

The climate change elements of the administration's Build Back Better bill, as proposed to Congress were more substantively and politically complicated, and they proposed more spending. The proposals were in some respects the centerpiece of the administration's wide ranging actions on the climate change agenda. The budget details of the act, as passed by the House of Representatives, are presented in Chap. 3. As of the end of mid-2022, passage of the bill by the Senate was still being prevented by two of the 50 Democratic Senators and all 50 of the Republican Senators.

2.6 Executive Order on Federal Government Sustainability [7]

On 8 December President Biden promulgated an unprecedented and extensive Executive Order with a target for the U.S. government's own emissions to be net zero by 2050, with an interim decrease of 65% by 2030. It also has a specific near-term goal of decreasing emissions in U.S. government buildings in 2032 by 50%. Because the government spends about $650 billion a year on purchases for its 300,000 buildings and 600,000 cars and trucks, the impact on emissions and production practices in the energy and transportation sectors could be substantial over time. Excerpts from the 34 sections of the order are available in Annex 1.

2.7 International Context

The increasing sense of urgency about climate change in many other countries led to additional international political pressure on the U.S. to be more responsive to the climate crisis, especially after four years of the Trump administration. The administration's responses began on 20 January with the announced re-entry into the Paris Agreement, and were followed by other policy changes during 2021 at international meetings.

2.7.1 The Paris Agreement

Rejoining the Paris Agreement and actively participating again in the annual Conferences of the Parties (COPs) of the UN Framework Convention on Climate Change (UNFCCC) are reminders that the international context of US policymaking is centrally important. Being back in the Paris Agreement changed the US status from being one of only a few countries in the world not belonging to being again one of the nearly 200 countries that do belong.

However, signing-up to the agreement is not the same as taking actions to achieve the non-binding Nationally Determined Contributions (NDCs) that all signatories pledged to do. Until the effective date of its formal withdrawal on 4 November 2020, the U.S. target was to reduce the level of U.S. greenhouse gas emissions by 26–28%

in 2025 compared with 2005. In fact, as of 2018, the U.S. reported a *3.7% increase* over 2005; by comparison, the EU reported a *25.2% decrease* [8]. As a result of its re-entry, the U.S. needed officially to reformulate its Paris Accord NDC, which it did by the time of the Leaders Summit.

2.7.2   Leaders' Summit [9]

The administration released its revised NDC in conjunction with the 22–23 April 2021 convening of an on-line international conference on climate change that it organized. The new target emission reduction was 50–52% by 2030 compared with 2005. Although there had already been a 13% reduction by 2019 over 2005, there was still a quite ambitious 37–39% reduction to be achieved within the decade. At that time, many of the details of the policies and paths by which such significant reductions could occur were still being decided. In any case, there would clearly have to be transformations of at least the electric power and transportation sectors, and those became center pieces of its plans, as envisaged in its *domestic* initiatives noted above.

2.7.3   G-20 Meeting [10]

The U.S. agreed in October 2021 at a G-20 meeting in Rome that it would stop government financing of 'new unabated coal power generation abroad'—as it had been doing through various international economic assistance and export promotion programs. At the same meeting, the administration made a commitment to resume contributions to a previously agreed collective international goal of $100 billion in annual support for developing countries' climate change programs.

2.7.4   COP26 in Glasgow [11–15]

An innovation at the Glasgow COP was a focus on sector-specific international agreements—including de-forestation and coal in particular. In addition, there was an international agreement on methane emissions, with a focus on the oil and gas in the energy sector. There were also revisions in international carbon market rules affecting offset programs, as well as individual updates for NDCs.

The deforestation agreement included a pledge to ban deforestation by 2030 [11]. The proposal was developed in part by the U.S. during 2021. It was signed by more than 100 countries, collectively with more than 85% of the world's forests. However, it was significantly weakened by Brazil's withdrawing its initial support, and the absence of explicit enforcement procedures. Its future effectiveness is thus in doubt, though its announcement at the COP meeting did create some sense of international progress on a particularly important and problematic sector.

The negotiations on the coal agreement [12] became contentious over the precise wording of a key provision, namely whether the signatories would agree to 'phase out' or 'phase down' their coal industries. As the clock was running out in the already overtime formal conclusion of the COP meetings, an agreement was reached when China and India's insistence on 'phase down' was accepted. Although the details of the U.S. role in the last-minute COP discussions was not publicly evident, for its own *domestic* political situation it was an acceptable compromise outcome on the international agreement. Indeed, since the U.S. administration was at the same time negotiating domestically with a Senator from the coal state of West Virginia over many elements of the Build Back Better Act [6], it was no doubt relieved that it did not have to defend an international agreement endorsing the 'phase out' of the coal industry. Nevertheless, as a signatory to the international agreement, the U.S. government did agree to end government financing of new unabated coal-fired electric power plants [13].

The U.S. was directly involved in developing the Global Methane Pledge with the European Union [14, 15]. The Pledge was signed by more than 100 governments and the European Commission, but not by China or Russia. The Pledge includes a 30% reduction in methane emissions by 2030. In terms of the U.S. government's climate change agenda, the international agreement is complementary to its own domestic initiative to reduce methane emissions, as embodied in the Build Back Better Act.

2.7.5   Special Report of the Intergovernmental Panel on Climate Change (IPCC)

A UN Intergovernmental Panel on Climate Change (IPCC) *Special Report on Global Warming of 1.5 °C* [16] that was released in 2021 focused attention on the significance of a difference of only 0.5 degrees in the previously agreed global warming targets and the increasingly urgent need for action. The study was undertaken in view of the Paris Agreement target to keep the global average temperature increase to 'well below 2 °C above pre-industrial levels' and to 'pursue efforts to limit the temperature increase to 1.5 °C above pre-industrial levels'. The IPCC special study found that the half degree Celsius difference has significant implications for climate policymaking—both in terms of differences in the policies and differences in their effects. In short, as the 'Summary for Policymakers' concluded: 'Climate-related risks to health, livelihoods, food security, water supply, human security, and economic growth are projected to increase with global warming of 1.5 °C and increase further with 2 °C' [17]. Thus, half a degree Celsius can make a big difference in the impacts of climate change—and hence climate change policies—to mitigate greenhouse gas and particulate emissions.

2.8 Domestic Institutional Constraints

For many kinds of policies, the President needs the support of majorities in both the House and the Senate, as well as key committee and sub-committee chairs and/or other party leaders. In the Senate, super-majorities of 60/100 are required for some issues. A policymaking process where both houses of Congress are centrally involved is the annual budgeting cycle. A noteworthy example was the President's statement in his 27 January Executive Order that he would 'eliminate fossil fuel subsidies from the budget request for Fiscal Year 2022 and thereafter,' as reported in Annex 1 of this chapter. The annual budget process is of course central to many climate change policy issues, but especially those involving large expenditures for energy, transportation and agricultural programs. The budget for FY 2022 was an especially significant and contentious one because it was Biden's first and contained numerous major changes in policy priorities, and it is analyzed in detail in Chap. 3.

Another issue requiring Congressional approval is the establishment of a national emission trading system (ETS) or another method of pricing carbon. In 2009 the American Clean Energy and Security Act, which would have established a national ETS, passed the House of Representatives by a vote of 255–176. A similar bill was proposed in the Senate by then Senator Kerry and others. Because of the Senate requirement for a 60 vote 'super' majority for cloture of debate, which was deemed unattainable, Senate Democratic leaders and President Obama decided not to press ahead for a vote in 2010 [18]. A decade later, a central climate change policy issue of the newly-elected Biden administration was whether and how to design an emission trading system—or an alternative tax system or fee-and-dividend system—that would create incentives to reduce emissions of greenhouse gases and particulate matter.

There are independent regulatory agencies that can create and abolish regulations for sectors of the U.S. economy without necessarily needing formal agreement by the President. The Security and Exchange Commission (SEC) is one that has received much attention by the Biden administration [19]. Some of the regulated sectors are significant sources of greenhouse gas and particulate emissions—such as energy and transportation, as discussed in Chap. 5. The President's ability to make climate policies is thus constrained by the action—or inaction—of regulatory agencies. Newly appointed members of various regulatory agency commissions in 2021 began to change the policies of the previous administration, but often encountered opposition from interested industries and members of the House and Senate.

Federal government executive actions, regulatory agencies' decisions, and congressional legislation can be over-ruled by federal courts at all levels—District Courts, Appeals Courts and the Supreme Court. During the four years of the Trump administration, a large number of their climate policy decisions were overturned by federal courts, including some by the Supreme Court. The fates in the courts of most cases were still pending at the end of 2021 [20].

2.9 Public Opinion

The new administration observed in its 27 January 2021 Executive Order that 'the attitude of the American people toward greater impetus … on climate change and doing something about it has increased across the board—Democrat, Republican, independent' [3]. There had indeed been significant shifts in public opinions over previous years—both in terms of perceptions of the problems and preferences about government policies. These shifts continued during 2021, particularly over the summer when there were many climate-related extreme weather events, as documented in the previous chapter.

Table 2.1 presents extensive data from repeated surveys in May and September 2021 that compare opinions on many dimensions before and after those extreme weather events [21]. There are also indicators in the table that nearly all of the distributions of the opinions in September 2021 reflected the highest levels of understanding and concern about climate change since 2008 when the long series of repeated surveys was first conducted.

Of course, there are not only differences over time in opinions but also differences among demographic groups. Table 2.2 compares opinions according to gender, age and education on key questions.

There are consistent patterns in the table. Females are more knowledgeable and more worried than males; similarly for young people versus older people; and similarly for college graduates versus those with high school or less education. The gap

Table 2.1 Opinions in the U.S. about climate change before and after the summer of 2021 [21]

Opinion	Percentage		
	May 2021	September 2021	Change
Think global warming is happening	70	76 [a]	+6
Are 'extremely' or 'very' sure global warming is happening	50	57 [a]	+7
Think global warming is caused mostly by human activities	57	60 [a]	+3
Think most scientists think global warming is happening	57	59 [a]	+2
Are 'very worried' or 'somewhat worried' about global warming	64	70 [a]	+6
Have 'personally experienced' the effects of global warming	42	52 [a]	+10
Think people in the U.S. are being harmed 'right now'	45	55 [a]	+10
'Often' or 'occasionally' discuss global warming with family/friends	33	39	+6
Global warming is personally important	67	71 [a]	+4
Global warming is affecting the weather 'a lot'	31	43 [a]	+12

[a] Highest since first survey in 2008

Table 2.2   U.S. Demographic group differences in public perceptions of climate change [22]

Issue	Percentage of demographic group that agrees					
	Gender		Age		Education	
	Females	Males	18–34	55+	College grade	High School or less
Most scientists believe global warming is occuring	73	63	74	62	80	58
Rising earth temperature is caused by human activities	72	56	80	54	72	60
Effects of global warming have already begun	66	51	72	51	70	47
Worry a great deal about global warming or climate change	52	34	59	36	46	43
Rising earth temperature is caused by human activities	72	56	80	54	72	60
Global warming will pose a serious threat to you or your way of life in your lifetime	51	35	61	29	47	44
Mean Difference	15>		23>		13>	

is biggest along age lines and smallest along educational lines. The gaps are most consistent on gender lines, where an average greater than 15% more females than males are knowledgeable and worried (with the smallest gap of 10% and a largest of 18%). A table in Annex 2 presents detailed comparisons among self-identified Democrats, Independents and Republicans in terms of the same questions.

In Table 2.3, there are international comparisons of the US and other countries. A key question is whether the US public tends to share perceptions of climate change issues with publics in other countries or to differ from them. The table is based on comparisons of public perceptions of 'major threats'—climate change among eight others. There is a tendency for the US public compared with the European countries in particular to be less concerned about climate change and more concerned about several other issues (infectious diseases, terrorism, cyberattacks and nuclear weapons). The US was an outlier—along with Australia— in its relatively low ranking of global warming as a 'major threat'. Since the corona virus pandemic was especially widespread in the US, it is perhaps not surprising that would be ranked first in the summer of 2020.

Table 2.3 Public perceptions of major threats: international comparisons of the US and 13 other countries [23]

Country	Question: For each of nine threats, which were the most frequently mentioned as a 'major threat to your country'	
	Percent saying 'global climate change'	'Global climate change' rank among 9 threats[a]
US	62	5
Australia	59	4
Belgium	70	1
Canada	67	1
Denmark	60	2
France	83	1
Germany	69	1
Italy	83	1
Japan	80	4
Netherlands	70	1
South Korea	81	4
Spain	83	1
Sweden	63	1
UK	71	3

[a] Pew survey of 14, 276 adults, 10 June to 3 August 2020

[b] The nine potential 'major threats' presented to the respondents were: global climate change, spread of infectious diseases, terrorism, cyber attacks from other countries, spread of nuclear weapons, condition of the global economy, global poverty, long-standing conflicts between countries or ethnic groups, large numbers of people moving from one country to another

2.10 Implications: Presidential Power and Its Limits

Shared authority among the three branches of the national government and between the national and state governments in a federal political system is an obvious basic feature of the institutional and legal context of U.S. policymaking on climate change issues. A president is always constrained in many ways by these institutional arrangements. Nevertheless, a determined and politically skillful president with hundreds of competent administrators in executive agencies and scores of advisors with specialized expertise can shape and push the agenda on specific issues.

Climate change issues, however, pose special challenges. One is that climate science is novel, complex and confusing to many non-experts among the public and in the congress at the national level and among state and local government officials. Although the public and officials alike are increasingly knowledgeable, climate change denial is still popular among some groups. There is also still much misunderstanding about the relationship between climate change and economics. Some people are still unaware of the economic costs imposed by climate change and thus the economic benefits of mitigating emissions and investing in adaptive measures. Moreover, technological innovations that can reduce emissions are often socially and economically disruptive.

Thus, a combination of scientific, economic and technological features of climate change issues make them politically challenging for any president who wants to put the country on more active policy paths. Such challenges—and opportunities—are abundant in the budget process analyzed in the next chapter.

Annex 1: President Biden's Executive Orders on: (A) Protecting Public Health and the Environment and Restoring Science to Tackle the Climate Crisis', (B) 'Tackling the Climate Crisis at Home and Abroad', and (C) 'Catalyzing Clean Energy Industries and Jobs Through Federal Sustainability'

(A)   *Excerpts from 'Executive Order on Protecting Public Health and the Environment and Restoring Science to Tackle the Climate Crisis'—signed on 20 February 2021* [2]

Sec. 1. Policy. ... It is ... the policy of my Administration to listen to the science; to improve public health and protect our environment; to ensure access to clean air and water; to limit exposure to dangerous chemicals and pesticides; to hold polluters accountable, including those who disproportionately harm communities of color and low-income communities; to reduce greenhouse gas emissions; to bolster resilience to the impacts of climate change; to restore and expand our national treasures and monuments; and to prioritize both environmental justice and the creation of the well-paying union jobs necessary to deliver on these goals.

Sec. 2. Immediate Review of Agency Actions Taken Between January 20, 2017, and January 20, 2021. ... The heads of all agencies shall immediately review all existing regulations, orders, guidance documents, policies, and any other similar agency actions ... promulgated, issued, or adopted between January 20, 2017, and January 20, 2021, that are or may be inconsistent with, or present obstacles to, the policy set forth in section 1 of this order. ... [T]he head of the relevant agency, as appropriate and consistent with applicable law, shall consider publishing for notice and comment a proposed rule suspending, revising, or rescinding the agency action within the time frame specified [for each of the following:] (i) Reducing Methane Emissions in the Oil and Gas Sector ... (ii) Establishing Ambitious, Job-Creating Fuel Economy Standards ... (iii) Job-Creating Appliance- and Building-Efficiency Standard ... (iv) Protecting Our Air from Harmful Pollution....

Sec. 3. Restoring National Monuments [i.e. specific areas of previously protected land: Grand Staircase-Escalante, Bears Ears, and Northeast Canyons and Seamounts Marine National Monuments. The Secretary of the Interior and the Attorney General are designated to determine whether and how these previously protected areas should be covered by further conservation measures.]

Sec 2.4. Arctic Refuge. …[P]lace a **temporary moratorium** on all activities of the Federal Government relating to the implementation of the Coastal Plain **Oil and Gas Leasing** Program [of 2020, which allowed oil and gas drilling in offshore areas in the Arctic National Wildlife Refuge.]

Sec 2.5. Accounting for the Benefits of Reducing Climate Pollution. (a) It is essential that agencies capture the full costs of greenhouse gas emissions as accurately as possible, including by taking global damages into account. Doing so facilitates sound decision-making, recognizes the breadth of climate impacts, and supports the international leadership of the United States on climate issues. The "social cost of carbon" (SCC), "social cost of nitrous oxide" (SCN), and "social cost of methane" (SCM) are estimates of the monetized damages associated with incremental increases in greenhouse gas emissions. They are intended to include changes in net agricultural productivity, human health, property damage from increased flood risk, and the value of ecosystem services. An accurate social cost is essential for agencies to accurately determine the social benefits of reducing greenhouse gas emissions when conducting cost–benefit analyses of regulatory and other actions. (b) There is hereby established an Interagency Working Group on the Social Cost of Greenhouse Gases (the "Working Group"). … The Working Group shall, as appropriate and consistent with applicable law: (A) publish an interim SCC, SCN, and SCM within 30 days of the date of this order, which agencies shall use when monetizing the value of changes in greenhouse gas emissions resulting from regulations and other relevant agency actions until final values are published; (B) publish a final SCC, SCN, and SCM by no later than January 2022.…

Sec 2.6. Revoking the March 2019 Permit for the Keystone XL Pipeline. … The Permit is hereby revoked in accordance with Article 1(1) of the Permit. … The Keystone XL pipeline disserves the U.S. national interest. The United States and the world face a climate crisis. That crisis must be met with action on a scale and at a speed commensurate with the need to avoid setting the world on a dangerous, potentially catastrophic, climate trajectory.

Sec 2.7. Other Revocations. [Ten Executive Orders, three Presidential Memoranda and two Guidance statements promulgated by the previous administration during 2016–2020 are revoked or subjected to review and revision].

(B)   *Excerpts from the 'Executive Order on Tackling the Climate Crisis at Home and Abroad'—Signed on 27 January 2021* [3]

[Themes]

- Putting the Climate Crisis at the Center of United States Foreign Policy and National Security
- Taking a Government-Wide Approach to the Climate Crisis
- Empowering Workers Through Rebuilding Our Infrastructure for A Sustainable Economy
- Empowering Workers by Advancing Conservation, Agriculture, and Reforestation
- Empowering Workers Through Revitalizing Energy Communities
- Securing Environmental Justice and Spurring Economic Opportunity.

[Examples of decisions and directions to agencies]

The United States will immediately begin the process of developing its nationally determined contribution [NDC] under the Paris Agreement.

… press for enhanced climate ambition and integration of climate considerations across a wide range of international fora, including the Group of Seven (G7), the Group of Twenty (G20), and fora that address clean energy, aviation, shipping, the Arctic, the ocean, sustainable development, migration, and other relevant topics.

… [D]evelop a climate finance plan … to assist developing countries in implementing ambitious emissions reduction measures, protecting critical ecosystems, building resilience against the impacts of climate change, and promoting the flow of capital toward climate-aligned investments and away from high-carbon investments.

…[P]repare … a National Intelligence Estimate on the national and economic security impacts of climate change.

… Use … all available [federal government] procurement authorities to achieve or facilitate: (i) a carbon pollution-free electricity sector no later than 2035; and (ii) clean and zero-emission vehicles for Federal, State, local, and Tribal government fleets, including vehicles of the United States Postal Service.

… [P]ause new oil and natural gas leases on public lands or in offshore waters pending completion of a comprehensive review and reconsideration of Federal oil and gas permitting and leasing practices…

…[E]liminate fossil fuel subsidies from the budget request for Fiscal Year 2022 and thereafter.

…[E]nsure that Federal infrastructure investment reduces climate pollution, and … require that Federal permitting decisions consider the effects of greenhouse gas emissions and climate change.

… [T]he Secretary of the Interior … shall submit a strategy to the [National Climate] Task Force within 90 days of the date of this order for creating a Civilian Climate Corps Initiative, within existing appropriations, to mobilize the next generation of conservation and resilience workers and maximize the creation of accessible training opportunities and good jobs. The initiative shall aim to conserve and restore public lands and waters, bolster community resilience, increase reforestation, increase carbon sequestration in the agricultural sector, protect biodiversity, improve access to recreation, and address the changing climate.

The Interagency Working Group [on Coal and Power Plant Communities and Economic Revitalization] shall coordinate the identification and delivery of Federal resources to revitalize the economies of coal, oil and gas, and power plant communities ....

...[E]nsure that Federal infrastructure investment reduces climate pollution, and ... require that Federal permitting decisions consider the effects of greenhouse gas emissions and climate change.

[Create a new White House Office, Assistant to the President, multi-agency Climate Task Force, White House Environmental Justice Interagency Council, and White House Environmental Justice Advisory Council].

- White House Office of Domestic Climate Policy, headed by an Assistant to the President—the National Climate Advisor
- National Climate Task Force, chaired by the National Climate Advisor, including the heads of 15 executive agencies, the Chair of the Council on Environmental Quality, the Director of the Office of Science and Technology Policy, and four advisors to the president on the White House staff
- White House Environmental Justice Interagency Council, chaired by the Chair of the Council on Environmental Quality
- White House Environmental Justice Advisory Council ..., which shall advise the Interagency Council and the Chair of the Council on Environmental Quality.

(C) Excerpts from the 'Executive Order on Catalyzing Clean Energy Industries and Jobs Through Federal Sustainability' [7]

...

Section 101. Policy.

... As the single largest land owner, energy consumer, and employer in the Nation, the Federal Government can catalyze private sector investment and expand the economy and American industry by transforming how we build, buy, and manage electricity, vehicles, buildings, and other operations to be clean and sustainable. ...

It is therefore the policy of my Administration for the Federal Government to lead by example in order to achieve a carbon pollution-free electricity sector by 2035 and net-zero emissions economy-wide by no later than 2050. Through a whole-of-government approach, we will demonstrate how innovation and environmental stewardship can protect our planet, safeguard Federal investments against the effects of climate change, respond to the needs of all of America's communities, and expand American technologies, industries, and jobs. ...

Sec. 102. Government-wide Goals.

... Through a coordinated whole-of-government approach, the Federal Government shall use its scale and procurement power to achieve:

(i)   100% carbon pollution-free electricity on a net annual basis by 2030, including 50% 24/7 carbon pollution-free electricity, as defined in section 603(a) of this order;

(ii)   100% zero-emission vehicle acquisitions by 2035, including 100% zero-emission light-duty vehicle acquisitions by 2027;

(iii)   a net-zero emissions building portfolio by 2045, including a 50% emissions reduction by 2032;

(iv)   a 65% reduction in scope 1 and 2 greenhouse gas emissions, as defined by the Federal Greenhouse Gas Accounting and Reporting Guidance, from Federal operations by 2030 from 2008 levels;

(v)   net-zero emissions from Federal procurement, including a Buy Clean policy to promote use of construction materials with lower embodied emissions;

(vi)   climate resilient infrastructure and operations; and.

(vii)   a climate- and sustainability-focused Federal workforce.

Sec. 201. Agency Goals and Targets. ...

Sec. 202. Reducing Agency Greenhouse Gas Emissions. ...

Sec. 203. Transitioning to 100% Carbon Pollution-Free Electricity. ...

Sec. 204. Transitioning to a Zero-Emission Fleet. ...

Sec. 205. Achieving Net-Zero Emissions Buildings, Campuses, and Installations. ...

Sec. 206. Increasing Energy and Water Efficiency. ...

Sec. 207. Reducing Waste and Pollution. ...

Sec. 208. Sustainable Acquisition and Procurement. ...

Sec. 301. Federal Supply Chain Sustainability. ...

Sec. 302. Supplier Emissions Tracking. ...

Sec. 303. Buy Clean. ...

Sec. 401. Engaging, Educating, and Training the Federal Workforce. ...

Sec. 402. Incorporating Environmental Justice. ...

Sec. 403. Accelerating Progress Through Public, Private, and Non-profit Sector Engagement. ...

Sec. 501. Establishment of the Office of the Federal Chief Sustainability Officer. ...

Sec. 502. Designation and Duties of Agency Chief Sustainability Officers. ...

Sec. 503. Agency Planning and Performance Management.

Sec. 504. Duties of the Chair of the Council on Environmental Quality. ...

Sec. 505. Duties of the Director of OMB. …

Sec. 506. Duties of the National Climate Advisor. …

Sec. 508. Establishment of Federal Leaders Working Groups. …

Sec. 509. Government-wide Support and Collaboration. …

Sec. 510. Additional Guidance and Instructions for Agencies. …

Sec. 511. Coordination of Administration Priorities. …

Sec. 601. Limitations. …

Sec. 602. Exemption Authority. …

Sec. 603. Definitions. …

Sec. 604. Revocation. Executive Order 13,834 of May 17, 2018 (Efficient Federal Operations), is revoked.

Sec. 605. Determination. …

Sec. 606. General Provisions. …

Annex 2: U.S. Public Opinion Survey Data: Detailed Analyses of Selected Issues

A.   Wording: 'Climate Change' or 'Global Warming'

Question wording matters, of course. But how much, about what, among whom? The results of a Pew Research Center report [24] from a split-halves analysis in a 2015 national survey are helpful; a randomly selected half of the respondents were asked if 'global warming should be a top priority for the president and congress' and the other half were asked if 'climate change should be a top priority for the president and congress'. The results were somewhat different for the two wordings: 9 percentage points more for 'global warming' among the Democrats and 3 points less for the Republicans. However, the basic patterns were the same: more Democrats than Republicans agreed by wide margins, regardless of the wording (Table 2.4).

Table 2.4 Comparison of results using 'Climate Change' or 'Global Warming' in survey questions [24]

Question wording	Percentage agreeing		
	Democrats/lean dem	Republicans/lean rep	Difference between parties
'Climate change'	46	19	27
'Global warming'	55	16	39
Difference in results	9	3	

Table 2.5 Basic beliefs: comparisons among democrats, independents and republications [25]

Issue	Percentage agreeing by party self-identification			
	Democrats	Independents	Republicans	Difference: dems. and reps
Most scientists believe global warming is occurring	87	67	44	43>
Rising earth temperature is caused by human activities	88	65	32	56>
Effects of global warming have already begun	82	59	29	53>
Worry a great deal about global warming or climate change	68	43	10	58 >
Global warming will pose a serious threat to you or your way of life in your lifetime	67	43	11	56>

B.   Party Differences in the U.S. on Key Dimensions of Attitudes

A Gallup poll in March 2021 [26] provides details about attitudinal differences among self-identified Democrats, Independents and Republicans, as in Table 2.5. The large differences between the two parties range from 43 to 58%.

Another survey taken in March 2021 by the Yale Program on Climate Change Communication and George Mason Center for Climate Change Communication [26] provides a detailed data set comparing registered American voters' attitudes on a series of key climate change issues. It is not only informative about the differences between the two parties, but also the factional splits within the parties. Results for one of the many questions are highlighted in Table 2.6. The patterns of inter-party differences are consistent with those in Table 2.5 above. The patterns of factional intra-party difference in Table 2.3 are also salient: among those identifying themselves as Democrats, there is a strong consensus with 97% of self-identified 'liberal' Democrats and 89% of 'moderate or conservative' Democrats wanting the US government to do 'more' or 'much more' … 'to address global warming'. In contrast to this 8% difference within the Democratic party, there was a 26% difference between the 'liberal/moderate' versus the 'conservative' Republicans.

Table 2.6 Core policy preferences: comparisons among democrats, independents and republicans [26]

Question and responses[a]	All Reg. voters (100%)	Total dems. (46%)	Liberal dems. (24%)	Mod./Cons. dems. (21%)	Ind./other (10%)	Total reps. (40%)	Lib./Mod. reps. (13%)	Cons. reps. (26%)
U.S. should do more or less to address global warming?								
Much more	27	47	54	38	18	8	14	4
More	38	46	43	51	39	25	36	20
More + much more	**65**	**93**	**97**	**89**	**57**	**33**	**50**	**24**
Currently right amount	18	5	2	7	19	34	35	34
Less + much less	**16**	**2**	**1**	**2**	**21**	**32**	**14**	**40**
Less	7	1	0	2	7	15	5	19
Much less	9	1	1	0	14	17	9	21

[a] Survey features: 922 registered voters. 18–29 March 2021. Totals for 'Democrats' and 'Republicans' include 'leaners' who said in follow-up to initial question whether they were 'closer to' one of the two parties. 'Independents' did not include 'leaners'

References

1. United Nations Framework Convention on Climate Change (UNFCCC). Time Series, Annex I. https://di.unfccc.int/time_series. Accessed January 26, 2021.
2. U.S. White House. *Executive Order on Protecting Public Health and the Environment and Restoring Science to Tackle the Climate Crisis.* January 20, 2021. https://www.whitehouse.gov/page/3/?s=executive+orders. Accessed 15 April 2021.
3. U.S. White House. *Executive Order on Tackling the Climate Crisis at Home and Abroad.* January 27, 2021. https://www.whitehouse.gov/briefing-room/presidential-actions/2021/01/27/executive-order-on-tackling-the-climate-crisis-at-home-and-abroad/. Accessed April 15, 2021.
4. U.S. White House. *Fact Sheet: The American Jobs Plan.* March 31, 2021. https://www.whitehouse.gov/briefing-room/statements-releases/2021/03/31/fact-sheet-the-american-jobs-plan/. Accessed April 15, 2021.
5. U.S. White House. *President Biden's Bipartisan Infrastructure Law.* https://www.whitehouse.gov/bipartisan-infrastructure-law/. Accessed January 30, 2022.
6. U.S. House of Representatives. *H.R.5376—Build Back Better Act 117th Congress (2021–2022).* https://www.congress.gov/bill/117th-congress/house-bill/5376/text. Accessed January 30, 2022.
7. U.S. White House. *Executive Order on Catalyzing Clean Energy Industries and Jobs Through Federal Sustainability.* December 8, 2021. https://www.whitehouse.gov/briefing-room/presidential-actions/2021/12/08/executive-order-on-catalyzing-clean-energy-industries-and-jobs-through-federal-sustainability/. Accessed December 14, 2021.
8. U.S. White House. (2015). *Fact Sheet: U.S. Reports its 2025 Emissions Target to the UNFCCC.* March 31, 2015. https://obamawhitehouse.archives.gov/the-press-office/2015/03/31/fact-sheet-us-reports-its-2025-emissions-target-unfccc. Accessed January 30, 2022.

 9. UN FCCC. (2022). *The United States of America Nationally Determined Contribution*. https://www4.unfccc.int/sites/ndcstaging/PublishedDocuments/United%20States%20of%20Amer ica%20First/United%20States%20NDC%20April%2021%202021%20Final.pdf. Accessed January 30, 2022.
10. U.S. White House. *Fact Sheet. United States Advances Shared Interests with G20 World Leaders and Delivers for the American People*. October 31, 2021. https://www.whitehouse. gov/briefing-room/statements-releases/2021/10/31/fact-sheet-united-states-advances-shared-interests-with-g20-world-leaders-and-delivers-for-the-american-people/. Accessed January 25, 2022.
11. Spring, J., & Jessop, S. (2021). Over 100 global leaders pledge to end deforestation by 2030. *Reuters*, November 3, 2021. https://www.reuters.com/business/environment/over-100-global-leaders-pledge-end-deforestation-by-2030-2021-11-01/. Accessed January 30, 2022.
12. UN FCCC. (2021). End of Coal in Sight at COP26. *External Press Release*, November 4, 2021. https://unfccc.int/news/end-of-coal-in-sight-at-cop26. Accessed January 30, 2022.
13. Volcovici, V. (2021). Biden orders U.S. to stop financing new carbon-intense projects abroad. *Reuters*, December 10, 2021. https://www.reuters.com/business/energy/biden-orders-us-stop-financing-carbon-intense-overseas-fuel-projects-2021-12-10/. Accessed January 30, 2022.
14. U.S. White House. (2021). *Fact Sheet: President Biden Tackles Methane Emissions*, November 2, 2021. https://www.whitehouse.gov/briefing-room/statements-releases/2021/11/ 02/fact-sheet-president-biden-tackles-methane-emissions-spurs-innovations-and-supports-sustainable-agriculture-to-build-a-clean-energy-economy-and-create-jobs/. Accessed January 30, 2022.
15. Global Methane Pledge. *About the Global Methane Pledge*. https://www.globalmethanepledge. org/. Accessed January 30, 2022.
16. Inter-governmental Panel on Climate Change (IPCC). (2021). *Special Report on Global Warming of 1.5°C*. https://www.ipcc.ch/sr15. Accessed December 1, 2021.
17. Inter-governmental Panel on Climate Change (IPCC). (2021). Summary for policymakers. *Special Report on Global Warming of 1.5°C*. https://www.ipcc.ch/sr15/chapter/spm/. Accessed December 1, 2021.
18. Brewer, T. (2015). *The United States in a warming world: The political economy of government, business, and public responses to climate change* (pp. 169–170). Cambridge University Press.
19. U.S. Securities and Exchange Commission (SEC). (2022). *SEC Response to Climate and ESG Risks and Opportunities*. https://www.sec.gov/sec-response-climate-and-esg-risks-and-opport unities. Accessed January 31, 2022.
20. Columbia Law School. Sabin Center for Climate Change Law. (2022). *U.S. Climate Change Litigation*. http://climatecasechart.com/climate-change-litigation/us-climate-change-litigation/. Accessed January 31, 2022.
21. Leiserowitz, A., Maibach, E., Rosenthal, S., Kotcher, J., Carman, J., Neyens, L., Marlon, J. M., Lacroix, K., & Goldberg, M. (2021) Climate Change in the American Mind, September 2021. *Yale Program on Climate Change Communication and George Mason Center for Climate Change Communication*. https://climatecommunication.yale.edu/wp-content/uploads/ 2021/11/climate-change-american-mind-september-2021.pdf. Accessed December 20, 2021.
22. Saad, L. (2021). Are Americans Concerned About Global Warming? *Gallup*, October 5, 2021. https://news.gallup.com/poll/355427/americans-concerned-global-warming.aspx. Accessed November 20, 2021.
23. Poushter, J., & Huang, C. (2020). *Despite Pandemic, Many Europeans Still See Climate Change as Greatest Threat to Their Countries*. September 9, 2020. https://www.pewresearch. org/global/2020/09/09/despite-pandemic-many-europeans-still-see-climate-change-as-gre atest-threat-to-their-countries/. Accessed October 15, 2021.
24. Pew Research. U.S. Public Views on Climate and Energy. November 25, 2019. https://www. pewresearch.org/science/2019/11/25/u-s-public-views-on-climate-and-energy/. Accessed September 25, 2021.
25. Gallup News Service. Environment. March 1–15, 2021. file:///D:/US%20CHAPS/210323 Satisfaction.pdf. Accessed December 22, 2021.

26. Leiserowitz, A., Maibach, E., Rosenthal, S., Kotcher, J., Carman, J., Wang, X., Goldberg, M., Lacroix, K., & Marlon, J. (2021). *Public support for international climate action*, March 2021. Yale Program on Climate Change Communication and George Mason Center for Climate Change Communication. https://climatecommunication.yale.edu/publications/public-support-for-international-climate-action-march-2021/. Accessed April 20, 2021.

Chapter 3
Who Gets What in the Budget

3.1 Overview of the Budget Process

The U.S. national government's budgeting process is centrally important in climate change policymaking for many reasons. The budget determines how much economic support government science agencies and private companies receive for their research and development programs. Such choices are based in part on economic cost-effectiveness and cost–benefit calculations that compare alternative technologies for reducing emissions. The budgeting process also allocates funds to state and local governments to support mitigation and adaptation programs. The presence or absence of budget support for many climate change policies can have significant long-term economic consequences; for instance, budgeting support for construction projects that reduce coastal flooding from the rising sea levels and increasing severity of hurricanes can significantly reduce their economic costs over many years. The budgeting process is thus an inherently distributive process that determines who gets how much government economic support of what kinds.

At the same time, many climate change issues involve highly technical scientific and technological topics. In short, understanding climate change budget issues requires wide-ranging inter-disciplinary analysis.

This chapter analyzes the fiscal year 2022 budget cycle beginning October 2021 and ending September 2022. In addition to the proposals for the normal FY2022 cycle, the new administration's ambitious and wide-ranging initiatives also involved new budget issues. The following sections thus consider first the FY2022 proposals, including the request for the Federal Emergency Management Agency (FEMA), then the 'infrastructure' bill, and finally the 'Build Back Better' bill, with its many climate-related programs.

© The Author(s), under exclusive license to Springer Nature Switzerland AG 2022
T. L. Brewer, *Transforming U.S. Climate Change Policies*,
SpringerBriefs in Energy, https://doi.org/10.1007/978-3-030-99716-8_3

3.2   FY2022 Request to Congress

The administration's 'normal' FY2022 request to Congress included substantial funds for 'Tackling the Climate Crisis' in the amount of $36 billion, which was $14 billion (or 64%) more than the enacted amount of $22 billion for FY2021. (During the four years of the previous administration's budget proposals, Congress regularly enacted more than the administration's requests on climate issues.) Table 3.1 lists the four main categories and 13 subcategories in the budget submission to Congress in August 2021.

The relatively large numbers for 'Innovation and Science' are conspicuous. The $10 billion for 'clean energy technologies' was about 30% more than the amount enacted by Congress for FY2021 during the previous administration. The FY2022 request was widely distributed among the electric, transportation, buildings and industry sectors of the economy. See Table 3.2 and Box 3.1 for the Energy Department programs.

Box 3.1 Highlights of the Energy Department's Justification of Its FY2022 Request [2]

The Office of Energy Efficiency and Renewable Energy (EERE) in the Department of Energy is responsible for developing and implementing many of the programs that are encompassed by the FY2022 entry 'Spurs Innovation in Clean Energy Technologies,' with a request for $10 billion. Much of EERE's request for more than $4 billion is for programs that are motivated by such a goal. Among them are those below.

Sustainable Transportation programs support 'RDD&D efforts to decarbonize transportation across all modes to enable the following: vehicle electrification; commercially viable hydrogen fuel cell trucks; sustainable aviation fuel from biomass; and waste carbon resources and low-GHG options for off-road vehicles, rail, and maritime transport.'

Renewable Power programs support 'RDD&D efforts in solar, wind, water, and geothermal power to help reduce the costs and accelerate the use and integration of renewables, contributing to a reliable, secure, and resilient grid which, in turn, produces many thousands of good-paying jobs.'

Energy Efficiency programs support 'RDD&D focused on the resilience of homes and buildings and strengthening U.S. manufacturing competitiveness. Ongoing efforts include the deployment of commercially ready technologies or demonstrations, as well as the acceleration of innovation to help decarbonize

Table 3.1 The administration's FY2022 request for 'Tackling the Climate Crisis' [1]

Program themes	Amount requested ($ bil.)[a]	Percent of $36 bil. request
Building clean energy projects and investing in resilience	*$5.6 bil*	*15.6%*
Improves energy efficiency, safety, and resilience of low-income homes and public buildings	1.7	4.7
Creates good-paying jobs building clean energy projects	2.0	5.6
Invests in climate resilience and disaster planning	0.815	2.3
Helps tribal nations address the climate crisis	0.450	1.3
Increases demand for american made, zero-emission vehicles through federal procurement	0.600	1.7
Helping communities left behind	*$5.9 bil.*	*16.4%*
Makes the largest investment in environmental justice in history	1.4	3.9
Propels an effort to create 250,000 jobs remediating abandoned wells and mines	0.580	1.6
Creates jobs improving critical water infrastructure[b]	3.6[b]	10.0
Partners with rural America to grow rural economies and tackle rural poverty	0.300	0.8
Increasing Competitiveness through Invest. in Innovation and Sci.	*$22.0 bil.*	*61.1%*
Advances climate science and sustainability research	4.0	11.1
Spurs innovation in clean energy technologies	10.0	27.8
Drives breakthrough solutions in climate innovation	1.0	2.8
Expands observations, research, and climate services	7.0	19.4
Leading the world toward achieving paris agreement objectives	*$2.4 bil.*	*6.7%*

(continued)

Table 3.1 (continued)

Program themes	Amount requested ($ bil.)[a]	Percent of $36 bil. request
Supports global emissions reductions[c]	2.4	6.7

[a] The table reflects rounded amounts in the source

[b] The rationale for including this subcategory with the programs for 'Tackling the Climate Crisis' in the original document is not clear. The description in the original source refers to replacing 'lead service lines,' repairing 'septic systems,' and more generally improving drinking water and waste water infrastructure

[c] Included $1.2 billion contribution to the multilateral Green Climate Fund for economic assistance to developing countries (there were no US contributions for the four years FY2018-2021); $0.485 billion for other multilateral climate programs; and $0.7 billion for Department of State and the U.S. Agency for International Development (AID) for developing countries' programs for emission reduction, clean energy and adaptation

Table 3.2 Energy department FY2022 budget request [2]

Programs	FY2022 request ($ millions)	Percent change over FY2021 enacted
Sustainable transportation		
Vehicle technologies	595	+48.8
Bioenergy technologies	340	+33.3
Hydrogen and fuel cell technologies	197.5	+31.7
Renewable power		
Solar power technologies	386.6	+38.1
Wind power technologies	204.9	+86.2
Water power technologies	196.6	+31.0
Geothermal technologies	163.8	+54.5
Energy efficiency		
Advanced Manufacturing	550.5	+39.0
Federal Energy Management Program	438.2	+995.4
Building Technologies	382	+31.7

energy-intensive industries, sustainably strengthen the domestic supply chain for critical minerals, and increase energy efficiency and demand flexibility for the 125 million U.S. homes and cmmercial buildings.'

ARPA-C

There is also a new program 'Advanced Research Projects Agency-Climate (ARPA-C), with an initial budget request for $200 million. The following are 'illustrative examples' of the kinds of programs ARPA-C may develop:

- Climate sensors and monitoring for dramatically improved greenhouse gas (GHG) detection, climate analysis, and severe event prediction. Data tools that can assess quantities and permanence of GHGs stored in land, underground, or in oceans, as well as provide relevant regional and local information for adaptation and long-term planning.
- Carbon neutral/negative agricultural production and general land, freshwater, and ocean use (including various carbon sequestration technologies and/or albedo engineering).
- Prevention of GHG emissions from land sources (methane from warming permafrost, landfills, and other activities); new approaches to permafrost protection.
- Carbon neutral waste and recycling, including e-waste processes that concurrently provide critical materials for climate mitigation technologies.
- Resilient infrastructure to protect against climate related severe events, including roads/transit; coastal impacts; building technologies (including reducing heat island impacts), self-healing materials, and air quality systems; water supply and distribution; agriculture and related supply chains.

The National Oceanic and Atmospheric Administration (NOAA) is also regularly a recipient of large appropriations for its many large-scale research programs in the on-going climate science activities of the federal government. The administration proposed - technically 'requested' - $6.983 billion for FY2022 to 'expand observations, research, and climate services,' which was an increase of $1.544 billion (28%) compared with the previously enacted amount for FY 2021 [3] (see Annex 1 for details).

3.3   Federal Emergency Management Agency (FEMA) [4–6]

Climate change issues are central to the budget of the U.S. Federal Emergency Management Agency (FEMA). For purposes of understanding and assessing FEMA's many programs from a climate change perspective, it is useful to consider separately its National Flood Insurance Program (NFIP) first and then all other programs.

A notable feature of FEMA's budget presentation is that it is unusual in its format and content because it focuses on *new budget authority* rather than *outlays*. The distinction between them is important and sometimes confusing. Outlays refer to the amounts of money that can be—and should be—spent during a specified fiscal year. Authorized amounts refer to ceilings or maximum amounts that are legally established for one or more fiscal years—sometimes for several fiscal years in one piece of authorizing legislation. The authorized amounts can thus provide guidance for future years to facilitate longer-term planning by executive branch agencies and the congress. In full format presentations of annual budget data, both outlays and authorized amounts are displayed. Sometimes, however, only one or the other is displayed—single-year outlay amounts more often than authorized amounts, either single-year or multi-year.

In the case of the administration's FY2022 proposals for FEMA, FEMA's budget focuses on *authorized* amounts because of the inherent uncertainties about the number and severity of the natural disasters that will lead to government actions to draw upon FEMA's resources during the forthcoming fiscal year.

The administration's FY2022 request for FEMA's National Flood Insurance Program (NFIP) was $214.7 million, which was about $10 million more than the FY2021 enacted amount of $204.4 million [9]. This was only a small proportion of total FY2022 request for the agency of $28.4 billion. Since it is an insurance program, a particularly important number for NFIP funds is its 'total capacity to pay claims,' which is based on 'flood fund available resources' plus 'reserve fund available resources' plus 'remaining borrowing authority.' As of 31 March 2012, the total was $16.5 billion. A major concern when natural disasters are especially numerous and large, as they were of course in 2021, is whether the total is adequate. For the longer term, as extreme weather events become more numerous and more severe because of climate change, the policies and financial stability of the NFIP become more salient and controversial.

Thus, in September 2021, when the NFIP was due for reauthorization in conjunction with the congressional budgeting process for FY 2022, the program was under increased scrutiny. Created in 1968 as an insurance subsidy program for buildings in flood-prone areas, the program had been controversial – and accordingly revised— from time to time. It had become particularly popular among owners of second homes in hurricane-prone sections of the east coast of the US, and also when its rates were raised to reflect increasing risks and damage claims. Although the agency's 'total capacity to pay claims' was about $16 billion in early 2021, the increase in its 'case-work totals' from 122 in the first quarter of calendar year 2020 to 166 in the first quarter of calendar year 2021 reflected the new flooding conditions of the 2020–2021 period. The increase in floods was not only evident in the states from Florida to New York on the Atlantic coast, Texas, Louisiana and Florida on the Gulf of Mexico coast and California on the Pacific coast, it was also evident because of floods of the Mississippi and other rivers in many mid-west states.

3.4   Infrastructure Bill [7]

Because national government spending on 'infrastructure' has often received bi-partisan support in the past, the Biden administration chose that as a promising subject to begin to reduce the unusually contentious relationships between the parties following the previous four years of overtly hostile relations. Infrastructure spending has been politically popular with incumbents of both parties because it produces

Table 3.3 Amounts for Selected Climate-Related Items Programs in 2021 Infrastructure Legislation [7]

Item	Proposed by President (USD 100 billion/10 years)	Passed by Congress (USD 100 billion/10 years)
Electric vehicle charging stations	7.7	7.5
Electric buses and ferries	7.5	7.5
Cleaner electric grid	73	65
Reduce emissions near airports	25	25
Reduce emissions near ports	17	17
Railroads	80	66
Public transit	39	39

high-profile, tangible projects that create local employment and improve local roads and other public facilities.

Thus, at the beginning of April the administration announced its intention to submit a formal budget proposal to congress for a diverse package of projects. The proposals extended beyond the traditional notion of 'infrastructure' to include not only roads and bridges. After some reductions, the bill was supported by 215 of the 221 Democrats who voted in the House of Representatives and 13 of the 213 Republicans in the House who voted. In the Senate, all 50 of the Democrats and 19 of the 50 Republicans voted to approve it. It was signed by President Biden on 15 November 2021 (Table 3.3).

The record floods of 2021 were reminders of how economically damaging floods can be. As for the future, the map in Fig. 3.1 indicates the areas of the US that are most exposed to flooding as sea level rise and more severe hurricanes affect the Atlantic and Gulf coasts. As a result, virtually the entire Atlantic coast and Gulf coast are at risk; however, rising sea levels will also affect the California coasts around Los Angeles and San Francisco, as well as much of the Washington coast. In addition, there are many inland areas that are exposed to increasing rain falls and the associated flooding, including parts of Oregon, Alabama, Tennessee, West Virginia, Pennsylvania, Ohio, Illinois and other areas near major rivers of course. Among the most at-risk counties are several in Florida, Louisiana and Texas.

3.5 Build Back Better Bill [9]

The Build Back Better bill included the center piece of the administration's climate change program. It proposed tens of billions of dollars in appropriations over a ten-year period to support electrification of vehicles, improvement of railroad service,

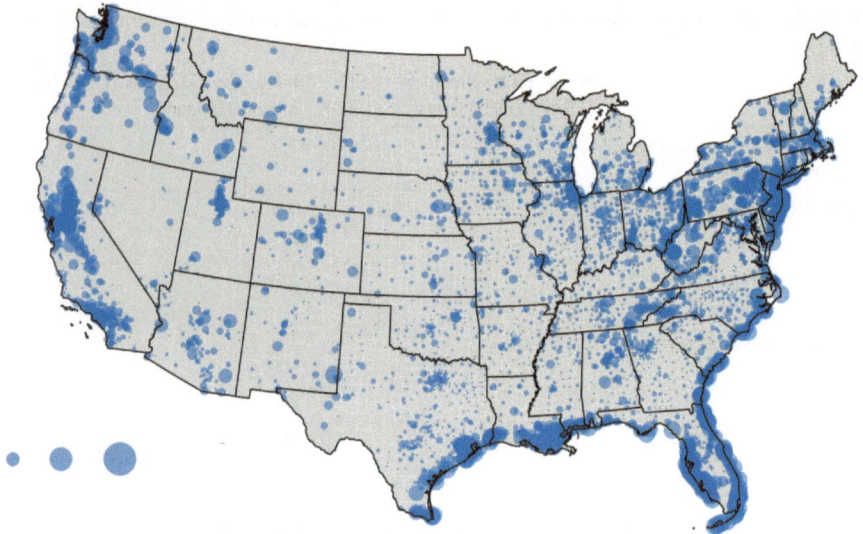

Fig. 3.1 Map of areas of the U.S that are most exposed to flooding [8]

reduction of emissions at airports and seaports and other transportation programs. It also provided for subsidies of improved residential insulation and significant increases for research and development of new nuclear technologies for electricity production as well as increased support for wind and solar energy projects, including offshore wind farms. See Annex 2 for an extensive list of the programs and budget amounts requested.

Along with many other programs that were focused on social programs that were not part of the climate change initiatives, the Build Back Better bill became stalled in the Senate after having been passed in the House of Representatives. The bill was opposed by all 50 of the Republicans in the Senate and two Democrats—enough to prevent its passage.

3.6 An Un-representative Senate

The impasse was in part the result of a well-known key institutional feature of the Senate that leads to an over-representation of fossil fuel interests in relatively small-population states [10]. The data in Table 3.4 reveal the extent to which small states with carbon intensive economies are over-represented by the constitutional provision that each state, regardless of its population size, is entitled to two senators. Thus, Wyoming, which ranks first as the most carbon-intensive state because of the prominence of coal and cattle in its economy, ranks last in terms of population size. Other

Table 3.4 Per capita carbon dioxide emissions of ten highest-ranking states [11]

State	Carbon Intensity[a] (CO_2 per capita)	Population[b] (millions)	Rankings among 50 states	
			Carbon Intensity rank	Population rank
Wyoming	112	0.6	1	50
North Dakota	75	0.8	2	47
West Virginia	52	1.8	3	39
Alaska	47	0.7	4	48
Louisiana	45	4.7	5	25
Montana	32	1.1	6	43
Kentucky	31	4.4	7	26
Indiana[c]	31	6.7	8	16
Nebraska	28	1.9	9	37
Oklahoma	27	3.9	10	27

[a] 2018 As indicated by CO_2 Carbon intensity can also be indicted by CO_2 e - according to which South Dakota (8th), Idaho (9th) and New Mexico (10th) are in the top ten, and Kentucky, Indiana and Oklahoma are not in the top ten

[b] 2018

[c] Indiana is a home to coal-fired power plants. Its agriculture industry is also exposed to billions of dollars in potential annual losses from increasingly frequent and severe droughts and floods as climatic conditions continue to change [12]

states with relatively carbon-intensive economies and relatively small populations are North Dakota, West Virginia and Alaska.

The contrast in terms of the relative degrees of states' carbon-intensity are yet more apparent in the data of Annex 4, where the ten states with the *least carbon-intensive* economies are indicated.

The year 2020 was not the first time that an over-represented group of carbon-intensive states have been able to thwart the climate policy preferences of the under-represented less carbon-intensive states. It happened in 2010 after the House of Representatives passed a bill that included the establishment of a nation-wide cap-and-trade system. Because of a group of Senators who opposed the legislation, the Obama administration decided to withdraw the proposal [10].

3.7 Implications for Climate Change Mitigation and Adaptation

Budgeting for climate change issues inevitably occurs in the context of interest in economic conditions—current and prospective. However, the economic issues tend to be different for mitigation and adaptation climate change policies. The economic benefits of *adapting* to the impacts of climate change tend to be short-term as well as long-term, and furthermore localized and readily apparent—and thus appealing to domestic constituencies. The economic benefits of *mitigating* emissions, in contrast, tend to be longer-term and more widespread among groups nationally and internationally—and therefore less compelling to many domestic constituents.

During the year 2021, these and other generalizations about the economics of climate change policymaking were affected by the extraordinary and obvious economic costs of the damage caused by climate-related droughts, fires and floods in the U.S., as documented in Chap. 1. Thus, the economic benefits of both mitigation and adaptation became more widely recognized, including in agricultural areas where there have been relatively high levels of climate change skepticism as well as coastal areas where sea level rise has already been widely recognized as an economic threat.

The economic consequences of the covid pandemic also affected the economics and politics of climate change policymaking. The combination of increased costs of government expenditures to address the public health crisis and reduced tax revenues because of lost jobs made increases in proposed government spending to meet the climate change challenges more politically problematic. In particular, some members of the Senate and House of Representatives were concerned about the overall government budget deficit.

The covid pandemic contributed to yet other economic conditions that in turn complicated the administration's climate policy initiatives. In particular, covid-induced interruptions in international and domestic supply chains and hence increased prices in fuels and food, for instance, complicated the administration's plans to reduce dependence on fossil fuels in transportation and electricity production. The administration was thus trying to induce significant changes in the U.S. economy at a time of increasing public concerns about prospective price levels and national economic growth.

Annex 1: Request for NOAA Appropriations for FY2022 [13]

The administration's Congressional submission for FY2022 for NOAA's climate programs included the following:

This budget supports NOAA's goal of scaling up efforts to research, mitigate, and adapt to the impacts of the climate crisis through investments in research, observations and forecasting, restoration and resilience, ecologically sound offshore wind development, and equity through programs that touch everyday lives. It also includes additional investments in fleet support, satellites, and space weather.

>$149,300,000 to strengthen core research capabilities to respond to increasing demand for the data, tools, and services that this research provides.

>$20,000,000 for climate-ready fisheries research that supports integrating climate science into fisheries assessments and management to address the impacts of climate change on fisheries, ecosystems, and communities.

>$368,195,000 to expand, renew, and improve our comprehensive environmental observing and forecasting systems to better support climate change-related decision-making. NOAA's ability to meet national weather research and forecasting needs depends on significant investments in global ocean observation systems, atmospheric observations, a seasonal forecast system, coastal ocean modeling, and aircraft observation capabilities.

>$17,000,000 to optimize and upgrade the Integrated Dissemination Program to improve capacity issues that will ensure reliable weather and climate predictions, forecasts, and warnings. Investments in Fire Weather will provide capacity for producing an integrated suite of fire prediction and decision-support tools to allow fire managers to better plan for and respond to wildfires.

>$5,000,000 to build a coastal inundation outlook capability at climate timescales to better support high-tide flood risk forecasting.

>$7,000,000 to accelerate nationwide improvement of the Flood Inundation Mapping program. Investments in ecological restoration and community resilience are integral to NOAA's climate strategy, and there is an increasing need for NOAA to create and foster natural and economic resilience along our coasts through our direct financial support, expertise, robust, on-the-ground partnerships, and place-based conservation activities.

>$259,330,000 to expand restoration and resilience efforts in ecosystems and communities, including through the National Coastal Resilience Fund and other grant programs, as well as through short-term employment and training opportunities that support the national Civilian Climate Corps.

>$23,500,000 in an initiative to assess place-based climate vulnerability to support engagement with local partners to strengthen conservation in existing national marine sanctuaries and marine national monuments (where NOAA is a co-manager), and assess new sites for sanctuary designation.

>$6,000,000 to foster ecological resilience by ensuring that coastal communities have access to NOAA observations and data products to inform conservation action and local management decisions. These collective restoration efforts will have multiple benefits, including buffering storms, reducing flooding, enhancing carbon sequestration, creating community-based jobs, and others.

>$20,380,000 to support NOAA's role in achieving the Administration's goal to deploy 30 GW of offshore wind in the U.S. by 2030, while protecting biodiversity and promoting ocean co-use.

>$57,900,000 to establish new programs focused on equity and environmental justice internally to build a more diverse and inclusive workforce, and externally to reach a broader range of Americans through service delivery and other outreach to underserved or disadvantaged communities.

>$9,000,000 to establish a new initiative to expand services across multiple programs and increase the number of storm-ready communities, including those rural and underserved communities that are especially vulnerable to extreme weather.

>$2,900,000 to accelerate progress on NOAA's diversity and inclusion implementation plan in support of identifying and addressing barriers to a diverse workforce to achieve a culture of equity and inclusion. NOAA's cutting-edge climate forecasting and service delivery, coupled with a robust approach to diversity, equity, and inclusion, make us well-positioned to make tangible improvements in the lives of those in the communities we serve, particularly those most vulnerable to climate change and its impacts.

>[S]ignificant investments for NOAA's observational infrastructure. The NOAA fleet and satellites are two key components of the NOAA mission. Fleet maintenance and construction are critical to NOAA's ability to collect climate data, and NOAA's satellites collect essential data that serve, in addition to being primary inputs for weather forecasting, as a long-term record for monitoring key climate parameters. There is increasing demand for NOAA's fleet and satellite systems to collect more accurate information and expand observing capacity. This request also supports additional capacity for the forecasting of space weather events, which can have far-reaching impacts on our Nation's economy, communications, and national security.

Annex 2: Build Back Better Budget Changes

See Table 3.5.

Annex 3: Calculations of the 'Social Costs of Carbon' in the Budget [14–19]

When the costs and benefits of climate-related programs in the US budget are estimated, a key input is the *social cost of carbon (SCC)*, which is an estimated US dollar cost of a metric tonne of carbon dioxide emissions. The number is used to compute the potential *benefits of reducing such emissions* through regulations of the sources of the emissions—or subsidies of alternatives with lower emissions. The costs of the regulations are estimated on the basis of industry economics and technologies.

The basic SCC concepts were developed in the US by economists in the 1980s and 1990s, beginning with Nordhaus in 1982 [16]. Over time, the key concepts have been refined, as have the data and models used to compute the numbers. At the same time, even as the core analyses and their results depended on economic reasoning and information technologies for processing the data, there have also been politically-driven changes in key numbers. Furthermore, there have been law suits about the appropriate institutionalized procedures for using the SCC computational results in the budgeting process.

Table 3.5 Proposed climate change appropriations in the build back better bill [9]

Item name[a] (section)	House amount and years
Clean heavy-duty vehicles—30101/132 (1) General	3.0 bil 10 years
Clean heavy-duty vehicles—132 (2) Non-attainment areas	2.0 bil 10 years
Clean heavy-duty vehicles—132 Reserve for EPA Administrative Costs	2% of 1–2 above
Reduce air pollution at ports—30102/133 (1) General Assistance	2.625 bil 6 years
Reduce air pollution at ports—133 (2) Nonattainment areas	0.875 bil 6 years
Greenhouse Gas Reduction Fund – 30103/134 (1) Zero-emission technologies	7.0 bil 3 years
Greenhouse gas reduction fund—134 (2) Zero-emission vehicle supply equipment	2.0 bil 3 years
Greenhouse Gas Reduction Fund – 134 (3) General assistance	11.97 bil 3 years
Greenhouse gas reduction fund—134 (4) Low-income and disadvantaged cities	8.0 bil 3 years
Greenhouse gas reduction fund—134 (5) Administrative costs	30.0 mil 10 years
Community wildfire air grants—30104	150.0 mil 10 years
Diesel emission reductions—30105	60 mil 10 years
Funding to address air pollution—30106 (1) Fenceline and screening air monitoring	117.5 mil 10 years
Funding to address air pollution—30106 (2) Multipollutant monitoring stations	50 mil 10 years
Funding to address air pollution—30106 (3) Air quality sensors in low-income …	3 mil 10 years
Funding to address air pollution—30106 (4) Emissions from wood heaters	15 mil 10 years
Funding to address air pollution—30106 (5) Methane monitoring	20 mil 10 years
Funding to address air pollution—30106 (6) Clear air act grants	25 mil 10 years
Funding to address air pollution—30106 (7) Other activities	45 mil 10 years
Funding to address air pollution—30106 (8) Mobil source GHG and emission Stds	5 mil 10 years
Air pollution at schools—30107 (a) In general	37.5 mil 10 years

(continued)

Table 3.5 (continued)

Item name[a] (section)	House amount and years
Air pollution at schools—30107 (b) Technical assistance	12.5 mil 10 years
Low emissions electricity program—30108/135 (1) Consumer-related education	17 mil 10 years
Low emissions electricity program—135 (2) Low-income and disadvantaged families	17 mil 10 years
Low emissions electricity program—135 (3) Industry-related outreach, tech. assist	17 mil 10 years
Low emissions electricity program—135 (4) State and local governments	17 mil 10 years
Low emissions electricity program—135 (5) Assess emission reductions	1 mil 10 years
Low emissions electricity program—135 (6) Ensure reductions are achieved	18 mil 10 years
Low emissions electricity program—135 EPA administrative costs	2% of 1–6 above
Clean air act sec. 211(O)—30109 (a) Test and protocol development	5 mil 10 years
Clean air act sec. 211(O)—30109 (b) Investments in advanced biofuels	10 mil 10 years
American innovation and manu. act—30110 (1) General	20 mil 5 years
(2) Implementation and compliance tools	3.5 mil 5 years
(3) Competitive grants	15 mil 5 years
Enforcement tech. and public info.—30111 (a) Compliance monitoring	37 mil 10 years
Enforcement Tech. and Public Info. – 30111 (b) Communications with ICIS	7 mil 10 years
Enforcement Tech. and Public Info. – 30111 (c) Inspection software	6 mil 10 years
GHG corporate reporting—30112	5 mil 10 years
Environ. product declaration assist. – 30,113	250 mil 10 years
30114 methane ??????????!!!!!!!!!!!!!!!!!!!!	?????
EPA Office of the Inspector General—30115	50 mil 10 years
Climate pollution reduct. grants—30116/137 (1) Planning grants	250 mil 10 years

(continued)

Table 3.5 (continued)

Item name[a] (section)	House amount and years
Climate pollution reduct. Grants—30116/137 (2) Implementation Grants	4.750 bil 5 years
EPA reviews—30117	20 mil 5 years
Low-embodied carbon labelling for construction mats. & trans. Projects—30118	100 mil 10 years
30119 ????????????????????	
30120 ????????????????????	
Grants to reduce com. Waste—30201/7011 (1) Organics recycling and food waste	95 mil 10 years
Grants to reduce com. Waste—30201/7011 (2) Other waste reduction activities	95 mil 10 years
Environ. and climate justice grants—30202/138 (1) Grants	2.8 bil 5 years
Environ. and climate justice grants—30,202/138 (2) Technical assistance for grants in (1)	200 mil 5 years
Data collection on natl. recycling—30203	10 mil 10 years
Lead remediation projects—30301	9 bil 5 years
Water assistance program—30302	225 mil 1 year
Subtitle D – Energy	
Part 1—Residential Eff. and Elect. Rebates	
Home energy rebates and train. grants—30411 (a) Home on-line training grants	360 mil 9 years
Home energy rebates and train. grants – 30411 (b) Home owner rebates	5.896 bil 9 years
Part 2—Building Efficiency and Resilience	
Critical facility modernization—30421	500 mil 10 years
Assist. for latest & zero building energy code Adoption … to carry out activities under Energy Policy and Conservation Act … (b) 30422	100 mil 10 years
Assist. for Latest & Zero Building Energy Code Adoption …to carry out activities under Energy Policy and Conservation Act … (c) 30,422	200 mil 10 years
Part 3—Zero-emissions vehicle infrastructure	

(continued)

Table 3.5 (continued)

Item name[a] (section)	House amount and years
Zero-emissions vehicle infrastructure grants—30431; Fin. Assist. to states for … (1) Level 2 elec. vehicle supply equipment	600 mil 7 years
(2) EV direct current fast charge equipment	200 mil 7 years
(2) Hydrogen fueling station	200 mil 7 years
Part 4–DOE Loan and Grant Programs	
(a) Commitment Authority—30441 DOE Loan Programs Office for guarantees under Energy Policy Act of 2005, sec. 1703	40 bil 5 years
(b) Guarantees under Energy Policy Act of 2005, sec. 1703	3.6 bil 5 years
Advanced tech. vehicle manufacturing—30442	3 bil 7 years
Domestic manufacturing conversion grants for plug-in elec. hybrid, elec. drive, and hydrogen fuel cell vehicles—30443	3.5 bil 1 year
Energy community reinvestment financing—30444/1706	5 bil 5 years
Tribal energy loan guarantees—30445 (a) Carry out energy policy act of 1992, sec. 2602(c)	200 mil 7 years
Tribal energy loan guarantees—30445 (c) Energy policy act of 1992, sec. 2602(c)	20 bil (years n.a.)
Part 5—Electric transmission	
Transmission line and Interline Incentives—30451 Grants and administrative expenses	1.5 bil 9 years
Transmission line and Interline incentives—30451 Direct Loans	500 mil 10 years
Grants to facilitate siting of interstate transmission lines—30452	800 mil 8 years
Wholesale electricity market Tech. Assist—30553	40 mil 10 years
Interregional and offshore wind electricity transmission planning, modeling, analysis—30454	100 mil 10 years
Part 6—Environmental reviews	
Department of Energy—30461	125 mil 10 years
Fedl. Energy Regulatory Commission—30462	75 mil 10 years
Part 7—Industrial	

(continued)

Table 3.5 (continued)

Item name[a] (section)	House amount and years
Advanced industrial facilities deployment—30471	4 bil 5 years
Part 8—Other energy matters	
Oversigh—30481	5 mil 10 years
Energy information agency—30482	40 mil 10 years

[a] Names in original source have been shortened for many items

Economic benefit–cost analyses began in the federal government during the Reagan administration in the early 1980s, as a way to evaluate the economic effects of government regulations; however, there were no calculations about the costs of climate change until many years later. Then, in 2008, the Center for Biological Diversity won a court case against the government because the government was not taking into account the economic costs of climate change in its assessments of programs and policies. Thus, in 2009, the federal government began to use SCC methods in the Department of Energy, and an Interagency Working Group developed standardized SCC methods in technical support documents in 2010, 2013, 2015 and 2016 [15].

President Trump abolished the Interagency Working Group in 2017, and there were also significant changes in SCC calculations at that time. For instance, the Obama administration had previously estimated the SCC for an electric power plant at $45 per tonne in 2020, while the Trump administration estimated it to be between $1 and $6. Thus, the benefit–cost calculations for the Clean Power Plan that had been passed during the Obama administration were transformed from being clearly favorable to the plan to being clearly unfavorable.

These and other issues about the SCC have been reviewed by the independent National Academy of Sciences in 2017 [19] in the executive branch and in 2020 by the Government Accountability Office of the Congress [18]. They were under further professional review in the first year of the Biden administration by 'The Social Cost of Carbon Initiative' organized by Resources for the Future in 2021 [17].

Throughout these reviews and changes in applications, there have been four central elements in the SCC system: (1) predictions of future emissions, which depend on assumptions about future populations and economic growth; (2) mathematical models of climate change that include variables such as emission levels and temperatures and much more; (3) the effects of climate change on economic sectors such as transportation; and (4) the use of percentage discount rates to determine the 'present values' of future economic costs and benefits. Of course, inevitably therefore the results depend on assumptions and related calculations concerning a broad range of environmental, economic, demographic and other conditions.

Yet, many of the politicized and otherwise controversial issues are focused on two numbers: (a) Should the effects of climate change include only those in the US or

Table 3.6 Variations in estimated economic values imputed to a tonne of carbon dioxide [15]

Discount rate (percent)	Global scope (US$ per tonne)	US National scope (US$ per tonne)
2.5	75	10
3.0	50	7
5.0	14	2
7.0	5	1

those in the entire world? (b) What is an appropriate discount rate? The results of the two choices have profound impacts on the estimated economic values of the benefits and costs of policy choices,. See Table 3.6 below, where the range of estimated US dollars per tonne is from $1 to $75.

Opinions about what is the best value depend on philosophical views as well as technical ones. The philosophical/ethical issues concern space and time. Since greenhouse emissions and policies to limit them inherently and inevitably pose global commons issues, it can be argued that the 'global' column is the most defensible one. Yet, since the issue is being addressed in the context of this 'Brief' on US policymaking in particular, perhaps an exclusive US scope is justified? Perhaps it is appropriate to consider both?

What is the best discount rate? That depends on one's valuation of the benefits and costs to future generations versus those accruing to current generations. A common ethically-based answer is that the discount value should be small since there is no ethical principle that the lives of current generations are inherently more valuable than the lives of later generations. However, there is another dimension to the issue— namely the increasing uncertainty about forecasts of future conditions with increasing distances into the future. It can be argued—and it is a common practice among economists—to discount predicted future events because there is inherently greater uncertainty about them the further into the future they are.

These core issues pervade applications of the SCC approach to evaluating policies concerning climate change that have also been adopted by state and local governments in the US and by governments in other countries. The answers, of course, are culturally-dependent.

Annex 4: States' Carbon-Intensity and Representation in the Senate

See Table 3.7.

Table 3.7 Carbon intensity of U.S. states (CO_2e emissions per million USD million GDP) [20]

First quintile			Last quintile		
Rank[a]	State	Amount (tonnes per mil. USD GDP)	Rank[a]	State	Amount (tonnes per mil. USD GDP)
1	Wyoming	2588	41	Connecticut	142
2	North Dakota	1641	42	Mississippi	140
3	West Virginia	1393	43	Washington	125
4	Montana	1175	44	Massachusetts	124
5	Louisiana	897	45	New York	119
6	New Mexico	896	46	New Hampshire	108
7	South Dakota	817	47	Vermont	80
8	Nebraska	790	48	((Dist. of Col.))	30
9	Iowa	738	49	Iowa	13
10	Alaska	695	50	Maine	−0.04[b]

[a] 2018

[b] Negative because of extensive forests that are net sinks

References

1. U.S. Office of Management and Budget (OMB). (2021). Compiled by the author.
2. U.S. Department of Energy (DOE). FY 2022 Budget Justification. May 28, 2021. https://www.energy.gov/cfo/articles/fy-2022-budget-justification. Accessed October 15, 2021.
3. U.S. National Oceanic and Atmospheric Administration (NOAA) (2021)
4. U.S. Federal Emergency Management Agency (FEMA). (2021). *National Flood Insurance Program (NFIP)*. www.fema.gov/flood-insurance. Accessed September 4, 2021.
5. Johnson, B. (2021). DHS budget: FEMA funding request focused on climate resilience, incident response. Government Technology & Services Coalition (*GTSC*), Homeland Security Today, June 5, 2021. www.hstoday.com. Accessed September 4, 2021.
6. U.S. Department of Homeland Security (DHS). (2021). Budget-in-Brief, *Fiscal Year 2022*. www.dhs.gov. Accessed September 4, 2021.
7. U.S. White House. *President Biden's Bipartisan Infrastructure Law*. https://www.whitehouse.gov/bipartisan-infrastructure-law/. Accessed January 30, 2022.
8. U.S. Federal Emergency Management Agency (FEMA). *Floods and Maps*. https://www.fema.gov/flood-maps. Accessed January 12, 2022.
9. U.S. House of Representatives. H.R.5376—Build Back Better Act117th Congress (2021–2022). https://www.congress.gov/bill/117th-congress/house-bill/5376/text. Accessed January 30, 2022.
10. Brewer, T. (2015). *The United States in a warming world: The political economy of government, business, and public responses to climate change* (pp. 169–170). Cambridge University Press.
11. World Resources Institute Climate Data Explorer. http://cait.wri.org/. Accessed November 5, 2021.
12. Bowling, L. C., Cherkauer, K. A., Lee, C. I., et al. Agricultural impacts of climate change in Indiana and potential adaptations. *Climatic Change,* 163, 2005–2027 (2020). Published by Springer, the version of record is available at: https://doi.org/10.1007/s10584-020-02934-9. Accessed January 25, 2021.
13. United States, National Oceanic and Atmospheric Administration. (2021). *Budget Estimates, Fiscal Year 2022. Congressional Submission.* August 2021. https://www.noaa.gov/sites/default/files/2021-06/NOAA%20FY22%20CJ.pdf. Accessed August 27, 2021.

14. Cho, R. (2021). *Social cost of carbon: what is it, and why do we need to calculate it?* Columbia Climate School. State of the Planet. https://news.climate.columbia.edu/2021/04/01/social-cost-of-carbon/. Accessed June 10, 2021.
15. Rennert, K., & Kingdon, C. (2019). *Social Cost of Carbon 101.* https://www.rff.org/publicati ons/explainers/carbon-pricing-101/. Accessed June 11, 2021.
16. Nordhaus, W. (1982). How fast should we graze the global commons? *American Economic Review,* 72(2), 242–246. https://www.econpapers.repec.org. Accessed 27 October 2021.
17. Resources for the Future. (2021). *The Social Cost of Carbon Initiative.*https://www.rff.org/top ics/scc/social-cost-carbon-initiative/. Accessed June 11, 2021.
18. U.S. Government Accountability Office (GAO). (2020). *Social cost of carbon: identifying a federal entity to address the national academies' recommendations could strengthen regulatory analysis.* GAO-20-254. Washington, DC, GAO. https://www.gao.gov/products/gao-20-254. Accessed June 10, 2021
19. United States, National Academy of Sciences (NAS). (2017). *Valuing climate damages: Updating estimation of the social cost of carbon dioxide.* https://www.nap.edu/catalog/24651/ valuing-climate-damages-updating-estimation-of-the-social-cost-of. Accessed June 10, 2021.
20. World Resources Institute (WRI). (2021). *8 Charts to Understand us state greenhouse gas emissions.* Compiled by the author. https://www.wri.org/insights/8-charts-understand-us-state-greenhouse-gas-emissions. Accessed November 6, 2021.

Chapter 4
The Future

4.1 Reframing the Issues

A key feature of the administration´s approach to climate issues has been to reframe them along several dimensions. One dimension is to give more attention to *adaptation* issues—based on a recognition of the accumulated scientific evidence and the increasingly damaging effects of extreme weather events such as those during 2021. Government adaptation programs often involve expenditures for local-level construction projects such as bridges, sea walls and other flood-reducing systems that are tangible and popular. Such localized spending projects sometimes thereby gain the support of some local residents who do not recognize the reality of climate change, but who nevertheless support the spending programs for local economic projects. It should also be noted, however, that most members of the opposition party in Congress nevertheless voted against the administration's infrastructure bill in 2021.

Another significant re-framing of the issues has been to place more emphasis on the *economic opportunities* of actively addressing emissions through subsidies and regulations in order to facilitate a transformation of industries to become more technologically advanced and more competitive in the future. This is in contrast to a common focus on the short-term economic costs of emission limitation measures—a tendency which is still relevant to the cost–benefit calculations discussed in the previous chapter. An element of the longer-term focus on the economic benefits of current expenditures has been to note the growth in U.S. jobs in transportation and energy industries, such as electric vehicles as well as solar and wind power.

Yet another reframing of issues has focused on the *ethical and political issues in the unequal distribution of the effects of climate change, particularly the effects on economically disadvantaged communities*. This emphasis was evident in the executive orders, legislative proposals and budget requests discussed in Chaps. 2 and 3. The focus on the effects on communities implicitly has also begun to draw attention to a collective sense of non-economic communal losses—for instance, when entire towns are destroyed by a single wildfire that has become more destructive because

© The Author(s), under exclusive license to Springer Nature Switzerland AG 2022   57
T. L. Brewer, *Transforming U.S. Climate Change Policies*,
SpringerBriefs in Energy, https://doi.org/10.1007/978-3-030-99716-8_4

of climate change induced drought conditions. The extent to which such concerns can be or will be extended internationally remains to be seen.

The re-framing has also included a sometimes *explicit preference for U.S. manufactured goods*—for instance in the specifics in implementation of the political rhetoric associated with the Build Back Better Bill. Such rhetoric targeted for domestic audiences was reinforced in the administration's positions on some international climate agreements, even as it played active leadership roles on others.

4.2   The U.S. as an International Leader and Laggard

Countries can be international leaders on climate change issues by supporting international agreements and adopting national policies. The U.S. record during the first year of the Biden administration was mixed. The U.S. rejoined the Paris agreement and strengthened its Nationally Determined Contributions to achieve the agreement's objectives, took initiatives in the international Leaders' Summit and G-20 meetings, and advocated for agreements at COP26, as noted in Chap. 2. However, it also has not been supportive of a strong international coal agreement or methane agreement—both because of domestic political pressure from coal industry interests, especially in the Senate—also as noted in Chap. 2.

4.3   Pricing Carbon

As documented in Chaps. 2 and 3, the Biden administration's climate change policy-making has featured a wide agenda. However, there has been one issue area where the administration has been more quiescent—namely putting a price on carbon through emissions trading and/or other means.

Many state governments had already begun to take significant action on climate change issues before the presidential election of 2016, and they increased their efforts after that presidential election year as the Trump administration canceled and reversed many Obama administration climate initiatives. In fact, a regional emission trading system had already been established by a group of northeastern states. Known as the Regional Greenhouse Gas Initiative (RGGI), it is focused on reducing carbon dioxide emissions in the power sector. It is the first mandatory market-based program in the United States to reduce greenhouse gas emissions. Its members are the eleven states of Connecticut, Delaware, Maine, Maryland, Massachusetts, New Hampshire, New Jersey, New York, Rhode Island, Vermont, and Virginia [1]. Meanwhile on the west coast, California had already created a cap-and-trade program in 2013. At the outset, the program included six greenhouse gases from industrial and electricity-producing facilities. In 2015, transportation fuels and natural gas were added. While California's population and economy have grown, greenhouse emissions have declined. The target established by the state legislature is that the 2030 emission level will be 40% below the 1990 level [2].

At the national level in 2021, with Democrats having a majority in the House of Representatives and half of the Senate seats, a variety of carbon pricing proposals were introduced in the 117th Congress [3], and they were awaiting formal action as of the end of January 2022. In the previous four years, there had also been proposals but there was virtually no prospect for their passage or serious consideration [4].

The Biden administration, meanwhile, was reluctant to support cap-and-trade [5], particularly in light of the 50/50 partisan distribution in the Senate and the elections already on the horizon for 2023.

In any case, the future of climate change policies will of course depend in part on institutional constraints and public opinion.

4.4   Institutional Constraints

Attempts to anticipate the future of U.S. climate change policy need to take into account institutional features that have directly affected key policies. One feature is the electoral-college presidential election system that yielded two recent minority presidents in terms of the votes by individuals tabulated on the day of election—George W. Bush in 2000 and Donald Trump in 2016. Both withdrew the U.S. from participation in international climate change agreements—Bush from the Kyoto Protocol and Trump from the Paris Agreement.

Another institutional feature constraining climate policymaking is highlighted in the previous chapter, where over-representation of fossil fuel interests in the Senate is documented. In that case, the administration proposed climate change legislation that was embedded in the Build Back Better Bill (which also included a much larger package of domestic social programs). In a previous situation, in 2010, the Obama administration supported a bill that would have established a national cap-and-trade system. Although it was passed by the House of Representatives, it faced much opposition in the Senate, where over-represented fossil fuel interests could defeat it. Because its fate in the Senate was doomed to failure, it was withdrawn to avoid a politically damaging legislative defeat for the administration [6].

4.5   The Generation Gap in Public Opinion

The answers to many questions about the future of public opinion depend on generational differences—in particular whether there are significant differences among age groups, whether they portend future differences, and whether the differences will matter. There is ample data that provide empirically based answers to the first question; answers to the other two are inevitably more speculative.

Several surveys have found consistent patterns in the differences among age groups. Gallup conducted surveys with repeated questions in annual surveys during 2015 through 2018—with a total sample of 4103 respondents. Their conclusion was that 'the extent to which Americans take global warming seriously and worry about it differs markedly by age, with adults under age 35 typically much more engaged with the problem than those 55 and older' [7]. In those surveys, the biggest generational gap was visible in the belief that global warming will pose a serious threat in one's own lifetime. This clearly reflects the different timeframes involved for each age group; the older one gets, of course, the less time in one's lifetime for global warming's effects to be experienced. The second-largest age gap occurs in the belief that global warming is caused by human activities. Younger adults are also significantly more likely to think news reports on global warming underestimate the problem. They are more likely to believe there is a scientific consensus that global warming is occurring. Younger and older Americans come closest in agreement in their views that the effects of global warming have already begun, and in self-reports of understanding global warming.

As for the size and the political significance of age differences, it is age differences within the Republican Party that are especially interesting for their potential impact on the future of U.S. policies. The results of a Pew survey [8] taken in May 2021 revealed both the partisan differences and age differences in U.S. adults' attitudes. The numeric specifics of the differences are displayed in Table 4.1.

The survey results reaffirm often-observed partisan differences: There was a 51% difference between the 87% of Democrats and the 36% of Republicans. However, there were not only significant differences *between* the parties in the opinions of partisans of all ages but also much bigger age differences *within* the Republican

Table 4.1 Partisan and age differences in climate change attitudes - reducing effects of climate change should be a top priority [9]

Age in 2021	All adults[a]	18–24	25–40	41–56	57–75	Difference:
Generation	– [a]	Generation Z	Millenials	Generation X	Boomers	Gen. Z—boomers
Year Born	– [a]	1997–2012	1981–1996	1965–1980	1956–1964	
All adults (%)	64	67	71	63	57	10
Democrats (%)	87	82	85	87	89	7
Republicans (%)	36	49	48	37	26	23
Difference: Dem.-Rep. (%)	51	33	37	50	63	

[a] Breakdowns for the 'Silent' generation—born in 1928–1945 and aged 76–93 in 2021—are not included in the original source table. A discussion of ages and definitions of the commonly used generation names is in [10]

party than *within* the Democratic party. The contrast between the parties is evident in the dis-aggregation of the party totals in Table 4.1. There is virtually unanimous agreement across age groups of Democrats, but a substantial split among Republicans according to age. Whereas a majority of the Republicans in the two youngest generations agreed, less than a majority of those in the two oldest generations agreed.

An obvious implication of these *intra*-party gaps and *inter*-party gaps is that partisan patterns in Congressional and Presidential election outcomes in coming years will be among the direct determinants of the future of U.S. government policies. The transformational policy changes initiated in 2021 as a result of the 2020 presidential election—and the partisan conflicts as a result of the congressional elections the same year—could be harbingers of the future. In that context, key questions concern what will happen to the generational gaps within the Republican party. Will the majorities of those in the lower age groups continue to hold their relatively climate-action friendly attitudes as they become older and replace the current older generations of party leaders and voters? Or will they become less climate-action friendly?

Yet, another possibility is that they and forthcoming young generations will both be more climate-action friendly—as the impacts of climate change continue to worsen and their impacts become increasingly costly in economic and human terms. It is likely that as the patterns of climate change conditions themselves continue to change in the future, therefore, the patterns of climate attitudes will also concomitantly change.

The *impacts* of any such changes in opinions at the national level, will depend in part on the extent to which the tendencies of un-representative outcomes of electoral college votes and Senate votes recur, as they have been analyzed in previous chapters.

References

1. Regional Greenhouse Gas Initiative (RGGI). (2022). *About the Regional Greenhouse Gas Initiative.* https://www.rggi.org/sites/default/files/Uploads/Fact%20Sheets/RGGI_101_Factsheet.pdf. Accessed January 5, 2022.
2. US, Energy Information Agency. (2018). *Today in Energy. California plans to reduce greenhouse gas emissions 40% by 2030*, February 2, 2018. https://www.eia.gov/todayinenergy/detail.php?id=34792. Accessed December 2, 2021.
3. Ye, J. (2022). *Carbon Pricing Proposals in the 117th Congress*, June 2021. Center for Climate and Energy Solutions. https://www.c2es.org/wp-content/uploads/2021/06/carbon-pricing-proposals-in-the-117th-congress.pdf. Accessed January 27, 2022.
4. Ye, J. (2022). *Carbon Pricing Proposals in the 116th Congress*. September 2020. Center for Climate and Energy Solutions. https://www.c2es.org/document/carbon-pricing-proposals-in-the-116th-congress/. Accessed January 27, 2022.
5. McCormick, M. (2022). *Carbon price is missing from Biden's overhaul of climate policy*, April 25, 2021. https://www.ft.com/content/4ecd8ace-8c5f-40ad-b2fa-c3bf27c2c621. Accessed January 29, 2022.
6. Brewer, T. (2015). *The United States in a warming world: The political economy of government, business, and public responses to climate change* (pp. 169–170). Cambridge University Press.

7. Reinhart, R. J. (2018). *Global warming age gap: Younger Americans most worried*. May 11, 2018. https://news.gallup.com/poll/234314/global-warming-age-gap-younger-americans-worried.aspx. Accessed November 12, 2021.

8. Funk, C. (2021). *Key findings: How Americans' attitudes about climate change differ by generation, party and other factors*. May 26, 2021. https://www.pewresearch.org/fact-tank/2021/05/26/key-findings-how-americans-attitudes-about-climate-change-differ-by-generation-party-and-other-factors/. Accessed 20 December 2021.

9. Tyson, A., Kennedy, B., & Funk, C., Gen, Z. (2021). Millennials stand out for climate change activism, Social Media Engagement with Issue. May 26, 2021. https://www.pewresearch.org/science/2021/05/26/gen-z-millennials-stand-out-for-climate-change-activism-social-media-engagement-with-issue/. Accessed December 20, 2021.

10. Dimock, M. (2019). *Defining generations: Where Millennials end and Generation Z begins.* January 17, 2019.https://www.pewresearch.org/fact-tank/2019/01/17/where-millennials-end-and-generation-z-begins/. Accessed December 12, 2021.